Physical Science

DIRECTED READING WORKSHEETS

This book was printed with soy-based ink on acid-free recycled content paper, containing 10% POSTCONSUMER WASTE.

HOLT, RINEHART AND WINSTON

A Harcourt Classroom Education Company

Austin • New York • Orlando • Atlanta • San Francisco • Boston • Dallas • Toronto • London

Welcome!

Imagine that you have just entered a foreign land and culture. What better way to experience the unfamiliar territory than to find a knowledgeable guide? He or she could point out beautiful landscapes and historical landmarks while dazzling you with interesting tidbits about the region. Your guide could help you make the most of your visit and help make it a visit you'll remember.

Well you *have* just entered a foreign land! You've entered the land of *Holt Science and Technology: Physical Science.* To help you make the most of your journey, use this booklet as your personal guide. Your guide will help you focus your attention on interesting images and important scientific facts. Your guide will also offer tips to help you understand the local language and ask you questions along the way to make sure you don't miss anything.

So sit back and get ready to fully experience *Holt Science and Technology*! Don't worry, this guide knows the ropes—all you have to do is follow along!

Art Credits
All art, unless otherwise noted, by Holt, Rinehart and Winston.
All work, unless otherwise noted, contributed by Holt, Rinehart and Winston.
Abbreviated as follows: (t) top; (b) bottom; (l) left; (r) right; (c) center; (bkgd) background.
Front cover (owl), Kim Taylor/Bruce Coleman, Inc.; (bridge), Henry K. Kaiser/Leo de Wys; (dove), Stephen Dalton/Photo Researchers, Inc.

Printed in the United States of America

ISBN 0-03-055704-6   19 20 21 170 06 05

▪ CONTENTS ▪

DIRECTED READING WORKSHEET

The World of Physical Science

As you read Chapter 1, which begins on page 4 of your textbook, answer the following questions.

Would You Believe . . . ? (p. 4)

1. What did Leonardo da Vinci learn from studying sea gulls?

_____

_____

_____

2. How did James Czarnowski get his idea for the penguin boat, *Proteus*? Explain.

_____

_____

_____

_____

_____

What Do You Think? (p. 5)

Answer these questions in your ScienceLog now. Then later, you'll have a chance to revise your answers based on what you've learned.

Investigate! (p. 5)

3. What is the seemingly impossible problem in this activity?

_____

_____

_____

Section 1: Exploring Physical Science (p. 6)

4. Your reflection on the inside of a spoon is different from your reflection on the outside of a spoon. True or False? (Circle one.)

Chapter 1, continued

That's Science! (p. 6)

5. Which of the following activities are part of "doing science"? (Circle all that apply.)

   a. asking questions   **c.** making observations
   b. being curious   **d.** taking an opinion poll

6. Only scientists can do science. True or False? (Circle one.)

Matter + Energy → Physical Science (p. 7)

7. Physical science is the study of energy and the stuff that every

   thing is made of. True or False? (Circle one.)

8. What do air, a ball, and a cheetah have in common?

   _____

9. All matter has energy, even if it isn't moving. True or False? (Circle one.)

10. What is one question you will answer as you explore physical science?

    _____

    _____

11. _____ is often divided into the categories of chemistry and physics. Chemistry is the study of different

    forms of _____ and how they

    _____. Physics is the study of

    _____ and how it _____
    matter.

Looking at the chart on page 8, identify each of the following descriptions as belonging to chemistry or physics by writing *physics* or *chemistry* in the space provided.

12. _____ how a compass works

13. _____ why water boils at 100°C

14. _____ how chlorine and sodium combine to form table salt

15. _____ how gravity causes the motion that makes a roller-coaster ride so exciting

16. _____ why you see a rainbow after a rainstorm

Physical Science Is All Around You (p. 9)

Choose the term in Column B that best matches the topic of physical science in Column A, and write the corresponding letter in the space provided.

Column A	Column B
_____ **17.** waves, currents, and ocean water chemistry	**a.** ecology
_____ **18.** the transfer of energy in a food chain	**b.** botany
_____ **19.** weather patterns	**c.** oceanography
_____ **20.** how plants use carbon dioxide and water to make food	**d.** biology
_____ **21.** earthquake waves	**e.** astronomy
_____ **22.** the motion of galaxies in the universe	**f.** geology
_____ **23.** how the brain sends electrical impulses throughout the body	**g.** meteorology

Physical Science in Action (p. 10)

24. Which of the careers on page 10 sounds most interesting to you, and why?

_____

_____

_____

Review (p. 10)

Now that you've finished Section 1, review what you learned by answering the Review questions in your ScienceLog.

Section 2: Using the Scientific Method (p. 11)

1. Advancements in science are mostly due to luck.

True or False? (Circle one.)

What Is the Scientific Method? (p. 11)

2. A series of _____ that scientists use to

_____ questions and

_____ problems is called the scientific

method.

Chapter 1, continued

Complete questions 3–10 after reading pages 12–18 in the text.
Match each sentence in Column A with one of the steps in the scientific method in Column B, and write the corresponding letter in the space provided.

Column A	Column B
_____ **3.** I told my classmates that Kaboing! shoes do not help you jump higher and that regular sneakers work better.	**a.** Ask a question.
_____ **4.** I wanted to know, "Will wearing Kaboing! shoes help me jump higher?"	**b.** Form a hypothesis.
_____ **5.** I jumped five times in a pair of Kaboing! shoes and recorded the height each time. After resting for 5 minutes, I repeated the test wearing my sneakers.	**c.** Test the hypothesis. **d.** Analyze the results. **e.** Draw conclusions.
_____ **6.** I thought I'd jump higher in Kaboing! shoes than in my sneakers.	**f.** Communicate the results.
_____ **7.** I jumped higher in my sneakers than I did in Kaboing! shoes. Kaboing! shoes do not help me jump higher.	
_____ **8.** The average height for the five jumps in Kaboing! shoes was 35.5 cm. The average height for the five jumps in my sneakers was 36 cm. On average, I jumped half a centimeter higher in my sneakers than I did in Kaboing! shoes.	

9. Take another look at question 5. Why do you think the student jumped five times in each pair of shoes instead of just once?

_____

_____

_____

_____

_____

_____

_____

10. Also in question 5, the student mentioned resting between tests. Why do you think that would have been important to the experiment?

_____

_____

_____

Review (p. 14)

Now that you've finished the first part of Section 2, review what you learned by answering the Review questions in your ScienceLog.

11. Which of the following statements is ALWAYS true for scientific investigations?

    a. Scientists never have a clear idea of the problem they are trying to solve.

    b. Scientists keep testing the same hypothesis.

    c. Scientists take accurate measurements and accurately record data.

    d. Scientists follow the steps of the scientific method in the same order.

Building Scientific Knowledge (p. 18)

12. An idea that is supported by many tests and experiments can

become a _____ or a

_____ .

13. Which of the following is NOT true of a scientific theory?

    a. It unifies hypotheses and observations that have been supported by testing.

    b. It can predict an observation you might make in the future.

    c. It can be changed or replaced.

    d. It is a simple guess.

Mark each of the following statements *True* or *False*.

14. _____ You could be arrested if you break a scientific law.

15. _____ Scientific laws are determined by committee.

16. _____ Laws tell you *why* something happens, not *what* happens.

17. _____ A scientific law is a summary of many experimental results and observations.

CHAPTER 1

18. Scientifically speaking, why do you think Figure 13 illustrates the big bang as a theory, not as a law?

_____

_____

_____

_____

_____

_____

_____

_____

_____

Review (p. 19)

Now that you've finished Section 2, review what you learned by answering the Review questions in your ScienceLog.

Section 3: Using Models in Physical Science (p. 20)

1. What did the MIT engineers hope to gain from making a model?

_____

_____

_____

What Is a Model? (p. 20)

2. You can represent an _____ or

_____ by using a model.

3. Which of the following are ways to use models in science? (Circle all that apply.)

    a. looking at the tiny parts of a microscopic cell on a cell diagram

    b. launching a homemade rocket in your backyard

    c. observing how the parts of matter fit together without being able to see the tiny particles

    d. testing a new design for a building on a computer before spending money on construction

4. How do you think a model rocket might help you understand a real rocket?

_____

_____

_____

_____

Models Help You Visualize Information (p. 21)

5. In Figure 15, how does using the spring toy as a model help you understand the behavior of sound waves?

_____

_____

_____

_____

6. When we picture things in our minds, we are creating models.

True or False? (Circle one.)

Models Are Just the Right Size (p. 22)

Decide whether a useful model for each of the following would be *larger* or *smaller* than the actual object, and write the appropriate answer in the space provided.

7. Mount Everest _____

8. a skyscraper _____

9. a computer chip _____

10. the moon _____

Models Build Scientific Knowledge (p. 22)

11. Models can be used as _____ to illustrate

_____ and _____

investigations.

12. Place each of the following statements in the correct sequence to explain how engineers could use *Proteus* to develop a new technology by writing the appropriate number in the space provided.

_____ Build a full-size penguin boat.

_____ Discover what factors affect the model's efficiency.

_____ Conduct tests on the model.

13. Why do you think the model in Figure 19 might be useful for understanding atomic theory?

_____

_____

_____

_____

Models Can Save Time and Money (p. 23)

14. How can cyber crashes like the one in Figure 20 save time and money?

_____

_____

_____

_____

Review (p. 23)

Now that you've finished Section 3, review what you learned by answering the Review questions in your ScienceLog.

Section 4: Measurement and Safety in Physical Science (p. 24)

1. At one time, systems of measurement were based on objects that varied in size, such as body parts and grains of barley.

True or False? (Circle one.)

The International System of Units (p. 24)

2. Using SI units helps scientists _____ and

_____ their results and observations.

3. Which unit of measurement is most practical for measuring the crane in Figure 22?

    a. centimeters (cm)   **c.** meters (m)

    b. kilometers (km)   **d.** millimeters (mm)

4. Before you can determine how many lenses will fit into a moving crate, what information do you need?

_____

_____

_____

5. Which of the following is NOT a valid unit of measurement for volume?

    a. milliliter   **c.** microgram

    b. cubic centimeter   **d.** liter

6. Would you measure the mass of a car in milligrams? Why or why not? If not, which unit would you use?

_____

_____

_____

_____

7. The _____ is the SI unit for temperature, but scientists often use degrees Celsius.

Derived Quantities (p. 26)

8. List two kinds of derived quantities.

_____

_____

9. What formula would you use to find out how much carpet it would take to cover the floor of your classroom? Write the equation.

_____

_____

10. How could you calculate the mass per unit of volume of the gear in the right column of page 27?

_____

_____

_____

_____

Safety Rules! (p. 27)

Match the safety icon in Column B to the correct description in Column A, and write the corresponding letter in the space provided.

Column A		Column B
____ **11.** hand safety	**a.**	
____ **12.** sharp object	**b.**	
____ **13.** clothing protection	**c.**	
____ **14.** chemical safety	**d.**	
____ **15.** eye protection	**e.**	
____ **16.** electrical safety	**f.**	
____ **17.** plant safety	**g.**	
____ **18.** heating safety	**h.**	
____ **19.** animal safety	**i.**	

Review (p. 27)

Now that you've finished Section 4, review what you learned by answering the Review questions in your ScienceLog.

CHAPTER 2 DIRECTED READING WORKSHEET

The Properties of Matter

As you read Chapter 2, which begins on page 34 of your textbook, answer the following questions.

Imagine . . . (p. 34)

1. What are two ways to tell "fool's gold" from real gold?

____

____

____

____

____

____

____

What Do You Think? (p. 35)

Answer these questions in your ScienceLog now. Then later, you'll have a chance to revise your answers based on what you've learned.

Investigate! (p. 35)

2. What will you do in this activity?

____

____

____

Section 1: What Is Matter? (p. 36)

3. What do a human, hot soup, and a neon sign have in common?

____

____

4. Anything that has ____________________ and ____________________ is called matter.

Chapter 2, continued

Matter Has Volume (p. 36)

Mark each of the following statements *True* or *False*.

5. _____ An object's volume is the amount of space the object takes up.

6. _____ Things with volume can't share the same space at the same time.

7. _____ When you measure a volume of water in a graduated cylinder, you should look at the bottom of the meniscus.

8. _____ A liquid's volume is usually expressed in grams or milligrams.

9. The volume of solid objects is expressed in

   _____ units. One milliliter is equal to

   _____ .

10. What three dimensions do you need to find the volume of a rectangular solid object?

    _____

    _____

11. You can't use a ruler to measure a gas, and you can't pour it into a graduated cylinder. So how do you find its volume?

    _____

    _____

    _____

Matter Has Mass (p. 38)

12. List the following objects in order from the least mass to the greatest mass: an elephant, a hamster, a skyscraper, the moon.

    _____

13. The mass of an object should be constant, yet the mass of the puppy in Figure 5 is not constant. Explain.

    _____

    _____

    _____

    _____

Chapter 2, continued

The Difference Between Mass and Weight (p. 39)

14. Why does all matter experience gravity?

    a. All matter has volume.   **c.** All matter is constant.

    b. All matter has mass.   **d.** All matter is stable.

15. Look at Figure 6. As two objects get _____,
the force of gravity between them increases.
(closer together or farther apart)

16. Weight is a measure of the gravitational force exerted on an

object. True or False? (Circle one.)

17. A brick weighs less in space than it does on Earth. Why?

_____

_____

_____

_____

18. Why do people tend to confuse the terms *mass* and *weight*?
(Circle all that apply.)

    a. Both remain constant on Earth.

    b. People use the terms interchangeably.

    c. Mass is the same thing as weight.

    d. Mass is also dependent on gravity.

Measuring Mass and Weight (p. 41)

19. The unit for _____ is the kilogram. The

unit for _____ is the newton.

Mass Is a Measure of Inertia (p. 42)

20. How do mass and inertia make it easier to pick up an empty
juice bottle than a full juice bottle?

_____

_____

_____

_____

Review (p. 42)

Now that you've finished Section 1, review what you learned by
answering the Review questions in your ScienceLog.

Chapter 2, continued

Section 2: Describing Matter (p. 43)

1. In a game of 20 Questions, the most helpful questions you can

   ask are about the _____ of the object.

Physical Properties (p. 43)

2. Which physical properties is the person in the picture on page 43
   asking about?

   _____

   _____

   _____

Match each physical property in Column B to the correct phrase in
Column A, and write the corresponding letter in the appropriate
space. Use the table on page 44 to help you.

Column A	Column B
____ 3. Sand does not dissolve in water.	a. state
____ 4. Gold can be made into gold foil.	b. thermal conductivity
____ 5. Ice is the solid form of water.	c. solubility
____ 6. Copper can be drawn out into wire.	d. density
____ 7. A foam cup protects your hand from being burned by the hot chocolate the cup contains.	e. ductility
____ 8. Ice cubes float in a glass of water because of their mass per unit volume.	f. malleability

9. In the formula for density, D means _____ ,

   V stands for _____ , and m stands for

   _____ .

10. What are two reasons why density is a useful property for
    identifying substances?

    _____

    _____

    _____

    _____

    _____

---

11. Using the table on page 45, list the elements mercury, water, gold, and oxygen from the densest to the least dense.

_____

12. What will happen if you shake the jar in Figure 12? Explain.

_____

_____

_____

_____

_____

_____

_____

13. Is 40 mL of oil denser than 25 mL of vinegar? _____

14. Density is dependent on the amount of substance you have.

True or False? (Circle one.)

Review (p. 46)

Now that you've finished the first part of Section 2, review what you've learned by answering the Review questions in your ScienceLog.

Chemical Properties (p. 47)

15. Which of the following are true of chemical properties? (Circle all that apply.)

   a. They indicate a substance's ability to change identity.
   b. They indicate one substance's ability to react with another.
   c. They describe matter.
   d. They can be observed with your senses.

16. In Figure 13, why isn't there rust in the painted areas of the car?

_____

_____

_____

_____

_____

Physical vs. Chemical Properties (p. 48)

17. Characteristic properties help scientists to distinguish one

substance from another. True or False? (Circle one.)

18. Which of the following represents ONLY physical properties? The table on page 48 may help you.

a. flammable, dense, malleable
b. malleable, reactive, dense
c. powdery, dense, red
d. clear, grainy, nonflammable

Physical Changes Don't Form New Substances (p. 48)

19. What is a physical change?

_____

_____

20. If you make a physical change to a substance, the identity of the

substance changes. True or False? (Circle one.)

21. Most physical changes are _____ to undo. (easy or difficult)

Chemical Changes Form New Substances (p. 49)

22. A chemical _____ describes a substance's

ability to change. A chemical _____ is the
process a substance goes through to turn into another substance.

23. How do you know that baking a cake involves chemical changes?

_____

_____

_____

24. In Examples of Chemical Changes, the odor of sour milk indicates

a chemical change has taken place. True or False? (Circle one.)

25. Some chemical changes can be reversed with more chemical

changes. True or False? (Circle one)

Review (p. 51)

Now that you've finished Section 2, review what you learned by
answering the Review questions in your ScienceLog.

CHAPTER

3 DIRECTED READING WORKSHEET

States of Matter

As you read Chapter 3, which begins on page 58 of your textbook, answer the following questions.

Imagine . . . (p. 58)

1. Lightning sometimes leaves behind a strange calling card. What is it called and how is it formed?

_____

_____

_____

_____

2. How do glassmakers use a change of state to make light bulbs, windows, and bottles?

_____

_____

_____

What Do You Think? (p. 59)

Answer these questions in your ScienceLog now. Then later, you'll have a chance to revise your answers based on what you've learned.

Investigate! (p. 59)

3. The purpose of this activity is

    a. to disinfect your hand with alcohol.

    b. to observe a change of state of alcohol.

    c. to observe a change of state of water.

Section 1: Four States of Matter (p. 60)

4. Look at Figure 1. Which of the following states of matter does Hero's steam engine demonstrate? (Circle all that apply.)

    a. solid   **c.** gas

    b. liquid   **d.** plasma

Moving Particles Make Up All Matter (p. 60)

5. The speed of the particles and the strength of the attraction

between them determine the _____ of the

substance.

Chapter 3, continued

Match the state of matter in Column B with the description in
Column A, and write the corresponding letter in the appropriate space.

Column A	Column B
____ **6.** Particles have a strong attraction to each other.	**a.** solid
____ **7.** Particles move independently of each other.	**b.** liquid
____ **8.** Particles are able to slide past one another but do not move independently of each other.	**c.** gas
____ **9.** Particles vibrate in place.	
____ **10.** Particles move fast enough to overcome nearly all of the attraction between them.	

Solids Have Definite Shape and Volume (p. 61)

11. The ship in the bottle in Figure 3 is a solid. How can you tell?

_____

_____

_____

12. Particles that are arranged in a repeating pattern of rows form
amorphous solids. True or False? (Circle one.)

Liquids Change Shape but Not Volume (p. 62)

13. How do the particles of a liquid make it possible to pour juice
into a glass?

_____

_____

_____

14. What does Figure 6 show you about the properties of a liquid?

_____

_____

15. Liquids tend to form in spherical droplets because of

_____ tension.

16. Water has a lower _____ than honey.

Chapter 3, continued

Gases Change Both Shape and Volume (p. 63)

17. How is it possible for a cylinder of helium to fill 700 balloons?

_____

_____

_____

_____

_____

Gas Under Pressure (p. 64)

18. The amount of _____ exerted on a given area is called pressure.

Review (p. 64)

Now that you've finished the first part of Section 1, review what you learned by answering the Review questions in your ScienceLog.

Laws Describe Gas Behavior (p. 65)

19. The volume of a gas is always the volume of its container.

True or False? (Circle one.)

20. Boyle's law states that if you keep the temperature constant for a fixed amount of gas, a decrease in pressure means a(n)

_____ in the volume of the gas.

21. Weather balloons are only partially inflated before they're released into the atmosphere. Why is that?

_____

_____

_____

_____

22. _____ is demonstrated by putting a balloon in the freezer.

23. All of the following remain constant in Figure 11 EXCEPT

    a. the type of piston.   **c.** the volume of the gas.

    b. the amount of gas.   **d.** the pressure.

Plasmas (p. 67)

Mark each of the following statements *True* or *False*.

24. _____ Plasma is the most common state of matter in the universe.

25. _____ Plasmas are made up of broken apart particles.

26. _____ Plasmas have a definite shape and volume.

27. _____ Plasmas and gases conduct electric current.

28. _____ Plasmas are affected by magnetic fields.

29. Lightning and fire are examples of _____ plasmas.

30. The incredible light show in Figure 12 is caused by plasma. How?

_____

_____

_____

Review (p. 67)

Now that you've finished Section 1, review what you learned by answering the Review questions in your ScienceLog.

Section 2: Changes of State (p. 68)

1. When a substance changes from one _____ form to another, we say the substance has had a change of state.

2. List the five changes of state.

_____

_____

3. The identity of a substance changes during a change of state.

True or False? (Circle one.)

4. Temperature is the measure of the speed of particles.

True or False? (Circle one.)

5. Temperature is a transfer of energy. True or False? (Circle one.)

6. Which of the following contains particles with the most energy?

   a. steam   **c.** ice

   b. liquid water   **d.** freezing water

_____

Chapter 3, continued

_____

Melting: Solids to Liquids (p. 69)

7. Could you use gallium to make jewelry? Why or why not?

_____

_____

_____

8. Melting point is a characteristic property, because it is the same

for different amounts of the same substance. True or False?
(Circle one.)

Freezing: Liquids to Solids (p. 69)

9. A substance's freezing point is the temperature at which it

changes from a _____ to a

_____ .

10. What happens if energy is added or removed from the ice water
in Figure 15?

_____

_____

_____

11. Freezing is considered an exothermic change because

_____ is removed from the substance.

Vaporization: Liquids to Gases (p. 70)

Choose the term in Column B that best matches the description in
Column A, and write the corresponding letter in the space provided.

Column A	Column B
____ **12.** vaporization at the surface of a liquid below its boiling point	**a.** boiling point
____ **13.** the change of state from a liquid to a gas	**b.** vaporization
____ **14.** vaporization that occurs throughout a liquid	**c.** steam
____ **15.** the product of vaporization of liquid water	**d.** evaporation
____ **16.** temperature at which a liquid boils	**e.** boiling

Condensation: Gases to Liquids (p. 71)

Mark each of the following statements *True* or *False*.

17. _____ At a given pressure, the condensation point for a substance is the same as its melting point.

18. _____ For a substance to change from a gas to a liquid, particles must clump together.

19. _____ Condensation is an exothermic change.

Sublimation: Solids Directly to Gases (p. 72)

20. Solid carbon dioxide isn't ice. So why is it called "dry ice"?

_____

_____

_____

21. The change of state from a solid to a _____ is called sublimation. Energy is required for sublimation, making it an _____ change.

Comparing Changes of State (p. 72)

22. Look at the table on page 72. Which two changes of state occur at the same temperature?
  a. condensation and melting
  b. sublimation and freezing
  c. vaporization and condensation
  d. melting and vaporization

Temperature Change Versus Change of State (p. 73)

Mark each of the following statements *True* or *False*. Figure 19 may help you.

23. _____ The speed of the particles in a substance changes when the temperature changes.

24. _____ The temperature of a substance changes before the change of state is complete.

25. _____ Energy must be added to a substance to move its temperature from the melting point to the boiling point.

Review (p. 73)

Now that you've finished Section 2, review what you learned by answering the Review questions in your ScienceLog.

CHAPTER

4 DIRECTED READING WORKSHEET

Elements, Compounds, and Mixtures

As you read Chapter 4, which begins on page 80 of your textbook, answer the following questions.

This Really Happened! (p. 80)

1. How do scientists think that the composition of the *Titanic*'s hull caused the "unsinkable" ship to sink?

_____

_____

_____

_____

2. Why might it be important to learn about the properties of elements, mixtures, and compounds?

_____

_____

_____

What Do You Think? (p. 81)

Answer these questions in your ScienceLog now. Then later, you'll have a chance to revise your answers based on what you've learned.

Investigate! (p. 81)

3. What do you think will happen to the ink of the black marker in this activity?

_____

_____

_____

_____

Section 1: Elements (p. 82)

4. What physical changes can you make to a substance to determine if it's an element? (Circle all that apply.)

   a. crushing   **c.** filtering
   b. grinding   **d.** passing electric current

An Element Has Only One Type of Particle (p. 82)

5. A pure substance is a substance that contains only one type of

   particle. True or False? (Circle one.)

6. In Figure 2, what do the skillet and the meteorite have in common?

   _____

   _____

Every Element Has a Unique Set of Properties (p. 83)

7. Look at the properties listed below. Circle the characteristic properties of elements.

   size   melting point   density   shape

   mass   volume   color   surface area

   hardness   flammability   weight   reactivity with acid

8. Suppose you have a cube of nickel and a cube of cobalt, but you don't know which is which. How could you use the characteristic properties listed in Figure 3 to figure out which cube is nickel and which is cobalt?

   _____

   _____

   _____

   _____

9. Most elements found in nature are combined with other elements.

   True or False? (Circle one.)

10. Why is it possible for us to find elements like gold, copper, and neon uncombined in nature?

    _____

    _____

    _____

Chapter 4, continued

Elements Are Classified by Their Properties (p. 84)

11. What are some common properties that most terriers share?

_____

_____

_____

12. Which of the following is a property that nickel, iron, and cobalt DON'T share?

    a. shine

    b. poor conductivity of electric current

    c. good conductivity of thermal energy

    d. None of the above

13. All elements can be classified as metals, metalloids, or

nonmetals. True or False? (Circle one.)

Look at the chart on page 85. Match the categories of elements in Column B with the correct properties in Column A, and write the corresponding letter in the appropriate space. Categories may be used more than once.

Column A	Column B
_____ **14.** malleable	**a.** metalloids
_____ **15.** dull or shiny	**b.** nonmetals
_____ **16.** poor conductors	**c.** metals
_____ **17.** tend to be brittle and unmalleable as solids	
_____ **18.** always shiny	
_____ **19.** may become good conductors when combined with other elements	
_____ **20.** graphite in pencils	
_____ **21.** always dull	
_____ **22.** used in computer chips	
_____ **23.** ductile	

Review (p. 85)

Now that you've finished Section 1, review what you learned by answering the Review questions in your ScienceLog.

Section 2: Compounds (p. 86)

1. When two or more elements are chemically combined to form a new pure substance, we call that new substance a

   _____ .

2. A compound is different from the elements that reacted to

   form it. True or False? (Circle one.)

3. List three examples of compounds you encounter every day.

   _____

Elements Combine in a Definite Ratio to Form a Compound (p. 86)

4. Which of the following is NOT true about compounds?
   a. Compounds join in specific ratios according to their masses.
   b. Mass ratios can be written as a ratio or a fraction.
   c. Compounds are random combinations of elements.
   d. Different mass ratios mean different compounds.

Every Compound Has a Unique Set of Properties (p. 87)

Mark each of the following statements *True* or *False*.

5. _____ Each compound has its own physical properties.

6. _____ Compounds cannot be identified by their chemical properties.

7. _____ A compound has the same properties as the elements that form it.

8. Sodium and chlorine can be extremely dangerous in their elemental form. So how is it possible that we can eat them in a compound?

   _____

   _____

   _____

   _____

   _____

   _____

   _____

Compounds Can Be Broken Down into Simpler Substances (p. 88)

9. How does opening a can of soda create the "fizz" in the drink?

_____

_____

_____

_____

_____

10. A physical change is the only way to break down a compound.
True or False? (Circle one.)

11. Look at the Physics Connection. The chemical process used
to obtain industrial products, such as hydrogen peroxide, is

called _____ .

Compounds in Your World (p. 89)

12. Which of the following methods are used by living organisms
to obtain nitrogen, an element needed to make proteins?
(Circle all that apply.)

a. Bacteria on the roots of pea plants make compounds from
atmospheric hydrogen.
b. Plants use nitrogen compounds in the soil.
c. Animals digest plants or animals that have eaten plants.
d. Plants take in carbon dioxide to make sugar.

13. What do most fertilizers, food preservatives, and medicines have
in common?

_____

_____

_____

14. The compound _____ is broken down to
produce the element used in cans, airplanes, and building materials.

Review (p. 89)

Now that you've finished Section 2, review what you learned by
answering the Review questions in your ScienceLog.

▲▲ CHAPTER 4

▲

Chapter 4, continued

Section 3: Mixtures (p. 90)

1. A pizza is not a mixture. True or False? (Circle one.)

Properties of Mixtures (p. 90)

2. When two or more materials combine without reacting with each other, they form a mixture. True or False? (Circle one.)

3. How do the granite in Figure 11 and the pizza at the top of the page show you that the identity of a substance doesn't change in a mixture?

_____

_____

_____

_____

_____

4. Mixtures are separated through _____

changes. _____ must be broken down
through chemical changes.

Look at the pictures on page 91. Match the technique for separating a mixture in Column B with the substances in Column A, and write the corresponding letter in the appropriate space.

Column A	Column B
____ **5.** crude oil	**a.** distill the mixture
____ **6.** aluminum and iron	**b.** centrifuge the mixture
____ **7.** parts of the blood	**c.** filter the mixture
____ **8.** sulfur and water	**d.** pass a magnet over the mixture

9. Granite can be pink or black, depending on the ratio of feldspar,

mica, and quartz. True or False? (Circle one.)

Review (p. 92)

Now that you've finished the first part of Section 3, review what you learned by answering the Review questions in your ScienceLog.

Chapter 4, continued

Solutions (p. 92)

10. Which of the following is NOT true of solutions?

    a. They contain a dissolved substance called a solute.

    b. They are composed of two or more evenly distributed substances.

    c. They contain a substance called a solvent, in which another substance is dissolved.

    d. They appear to be more than one substance.

11. In a gaseous or liquid solution, the volume of solvent is

    _____ the volume of solute.

    (less than or greater than)

12. The solid solution used to build the ship *Titanic* was a(n)

    _____ called steel.

13. Which of the following is true of particles in solutions?

    a. Particles scatter light.

    b. Particles settle out.

    c. Particles can't be filtered out of their mixtures.

    d. Particles are large.

14. Concentration is a measure of the volume of a solution.

    True or False? (Circle one.)

15. What is the difference between a concentrated solution and a dilute solute?

    _____

    _____

16. When a solution is holding all the solute it can hold at a given

    temperature, we say the solution is _____ .

17. Solubility is not dependent on temperature. True or False? (Circle one.)

18. Solubility is measured in grams of solute per

    _____ of solvent.

19. Solubility of *gases* in liquids tends to _____

    with an increase in temperature. Solubility of *solids* in liquids

    tends to _____ with an increase in temperature.

▲ ▲ ▲ CHAPTER 4

20. What are three ways to make a sugar cube dissolve more quickly in water?

_____

_____

Suspensions (p. 96)

21. Which of the following does NOT describe a suspension?

    a. Particles are soluble.
    b. Particles settle out over time.
    c. Particles can be seen.
    d. Particles scatter light.

22. Look at the Life Science Connection at the top of the page. Why is blood a suspension?

_____

_____

Colloids (p. 97)

23. What do gelatin, milk, and stick deodorant have in common?

_____

Match the mixtures in Column B to the characteristics in Column A, and write the corresponding letter in the appropriate space. Mixtures may be used more than once.

Column A	Column B
____ **24.** Particles do not settle out.	**a.** colloids and suspensions
____ **25.** Particles are larger.	
____ **26.** Particles scatter a beam of light.	**b.** colloids and solutions
____ **27.** Particles cannot be filtered out.	

28. Look at Figure 18. How can a colloid be dangerous to drivers?

_____

_____

Review (p. 97)

Now that you've finished Section 3, review what you learned by answering the Review questions in your ScienceLog.

CHAPTER

5 DIRECTED READING WORKSHEET

Matter in Motion

As you read Chapter 5, which begins on page 106 of your textbook, answer the following questions.

Would You Believe . . . ? (p. 106)

1. One reason Native Americans played the game of lacrosse was for fun. What was the other reason?

_____

_____

2. What are the advantages of using a lacrosse stick? (Circle all that apply.)

   a. It makes it possible to throw the ball at speeds over 100 km/h.
   b. It helps the defense players hold back the offense.
   c. It makes it possible to throw the ball over 100 km.
   d. It protects the hand from injury from the high speed ball.

What Do You Think? (p. 107)

Answer these questions in your ScienceLog now. Then later, you'll have a chance to revise your answers based on what you've learned.

Investigate! (p. 107)

3. What do you predict this activity will demonstrate?

_____

_____

_____

Section 1: Measuring Motion (p. 108)

4. Name something in motion that you can't see moving.

_____

_____

Observing Motion (p. 108)

5. To determine if an object is in motion, compare its position over

   time to a _____ point.

_____

6. Buildings, trees, and mountains are all useful reference points. Why?

_____

_____

7. Can a moving object be used as a reference point? Explain.

_____

_____

_____

Speed Depends on Distance and Time (p. 109)

Each of the following statements is false. Change the underlined word to make the statement true. Write the new word in the space provided.

8. <u>Motion</u> is the rate at which an object moves.

_____

9. How fast an object moves depends on the distance traveled and the <u>road</u> taken to travel that distance.

_____

10. The SI unit for speed is <u>km/h</u>.

_____

11. Why is it useful to calculate average speed?

_____

_____

12. Write out in words how to calculate average speed.

_____

_____

_____

13. Look at the Brain Food on p. 109. Suppose a car travels 250 m in 10 seconds. Is its average speed greater than or less than that of a running cheetah?

_____

_____

Chapter 5, continued

Velocity: Direction Matters (p. 110)

14. Why don't the birds in the riddle end up at the same destination?

_____

_____

15. Velocity has speed and _____ .

16. Which of the following does NOT experience a change in velocity?

    a. A motorcyclist driving down a straight street applies the brakes.

    b. While maintaining the same speed and direction, an experimental car switches from gasoline to electric power.

    c. A baseball player running from first base to second base at 10 m/s comes to a stop in 1.5 seconds.

    d. A bus traveling at a constant speed turns a corner.

17. To find the resultant velocity, add velocities that are in

_____ direction(s). Subtract velocities

that are in _____ direction(s).

Review (p. 111)

Now that you've finished the first part of Section 1, review what you learned by answering the Review questions in your ScienceLog.

Acceleration: The Rate at Which Velocity Changes (p. 112)

18. Why did the neighbor say you had great acceleration as you slowed down and swerved to avoid hitting a rock?

_____

_____

_____

_____

_____

19. Write the formula for calculating acceleration in the space below.

20. Scientifically speaking, how do you know the cyclist in Figure 4 is accelerating?

_____

_____

_____

_____

_____

_____

21. Another name for acceleration in which velocity increases is

_____ acceleration.

22. Negative acceleration, or acceleration in which velocity

decreases, is also called _____ .

23. What kind of acceleration is occurring in Figure 5?

_____

24. When you are standing completely still at the equator, you are

accelerating. True or False? (Circle one.)

25. How can you recognize acceleration on a graph?
   a. The graph shows distance versus time.
   b. The graph shows time versus distance.
   c. Velocity increases as time passes.
   d. The graph is a straight line.

Review (p. 114)

Now that you've finished Section 1, review what you learned by answering the Review questions in your ScienceLog.

Section 2: What Is a Force? (p. 115)

Mark the following statements *True* or *False*.

1. _____   All forces have size and direction.

2. _____   A force is a push or a pull.

3. _____   Forces are measured in liters.

Chapter 5, continued

Forces Act on Objects (p. 115)

4. You can exert a push force without there being an object to

receive the force. True or False? (Circle one.)

5. Name three examples of objects that you exert forces on when
you are doing your schoolwork.

_____

_____

_____

_____

_____

6. In which of the following situations is a force being exerted?
(Circle all that apply.)

a. A woman pushes the elevator button.
b. A pile of soil sits on the ground.
c. Socks like the ones in Figure 7 cling together when they have
just come out of the dryer.
d. Magnets stick to the refrigerator.

Forces in Combination (p. 116)

7. In Figure 8, how does the net force help the students move the
piano?

_____

_____

_____

_____

8. Suppose the dog on the left in Figure 9 increased its force to
13 N. Which dog would win the tug-of-war? Explain.

_____

_____

_____

_____

Unbalanced and Balanced Forces (p. 117)

9. Why is it useful to know the net force?

_____

_____

10. Forces are unbalanced when the net force is not equal to

_____ .

11. To start or change the motion of an object, you need a(n)

_____ force. (balanced or unbalanced)

12. Forces are balanced when the net force applied to an object is

_____ zero.
(less than, greater than, or equal to)

13. Are the forces on the cards in Figure 10 balanced? How do you know?

_____

_____

Review (p. 118)

Now that you've finished Section 2, review what you learned by answering the Review questions in your ScienceLog.

Section 3: Friction: A Force that Opposes Motion (p. 119)

1. What force is responsible for the painful difference between sliding on grass and sliding on pavement?

_____

_____

The Source of Friction (p. 119)

2. Friction occurs when the hills and valleys of two surfaces stick together. True or False? (Circle one.)

3. Pavement creates more friction than grass. Why is that?

_____

_____

_____

Chapter 5, continued

_____

4. Why is more force needed to slide the larger book in Figure 12?

_____

_____

_____

_____

5. Friction is affected by the amount of surface that is touching.

True or False? (Circle one.)

Types of Friction (p. 121)

Match each type of friction in Column B with its example in
Column A, and write the corresponding letter in the space provided.

Column A	Column B
____ **6.** a hockey puck crossing an ice rink	**a.** sliding friction
____ **7.** a crate resting on a loading ramp	**b.** rolling friction
____ **8.** wheeled cart being pushed	**c.** fluid friction
____ **9.** air rushing past a speeding car	**d.** static friction

10. Static friction is at work if you try to drag a heavy suitcase along

the floor and the suitcase _____.
(moves or doesn't move)

11. As soon as an object starts moving, static friction

_____. (increases or disappears)

Friction Can Be Harmful or Helpful (p. 123)

12. How does friction harm the engine of a car?

_____

_____

_____

13. Why do you need friction to play sports?

_____

_____

14. Which of the following are ways to reduce friction?
(Circle all that apply.)

    a. Use a lubricant.

    b. Make rubbing surfaces smoother.

    c. Push surfaces together.

    d. Change sliding friction to rolling friction.

Review (p. 124)

Now that you've finished Section 3, review what you learned by answering the Review questions in your ScienceLog.

Section 4: Gravity: A Force of Attraction (p. 125)

1. Why did the astronauts in Figure 18 bounce on the moon?

_____

_____

2. The force of attraction between two objects due to their masses

is the force of _____ .

All Matter Is Affected by Gravity (p. 125)

3. Does all matter experience gravity? Explain.

_____

_____

_____

4. The force that pulls you toward your pencil is called the

_____ force.

5. Look at the Life Science Connection. Scientists think seeds can

sense gravity. True or False? (Circle one.)

6. Since all objects are attracted to each other due to gravity, why
can't you see the objects moving toward each other?

_____

_____

_____

_____

Chapter 5, continued

7. How are objects around us affected by the mass of the Earth?

_____

_____

_____

_____

The Law of Universal Gravitation (p. 126)

8. What did Newton figure out about the moon and a falling apple?

_____

_____

_____

_____

9. Newton's law of universal gravitation describes the relationships between all of the following EXCEPT

    a. distance.   **c.** heat.

    b. mass.   **d.** gravitational force.

10. Which of the following objects are subject to the law of universal gravitation? (Circle all that apply.)

    a. satellites   **c.** frogs

    b. water   **d.** stars

11. If the distance between the objects are the same, the gravitational force between two feathers is greater than the gravitational

force between two bowling balls. True or False? (Circle one.)

12. If two objects are moved _____ each other, the gravitational force between them increases. (away from or toward)

13. What is the mathematical expression for the law of universal gravitation?

14. Read the Astronomy Connection on page 127. In a

_____ , gravity is so great that even light can't escape.

15. Why doesn't the sun's gravitational force pull you off the Earth?

_____

_____

_____

16. What would happen to the Earth and other planets in the solar system without the sun's gravitational force?

_____

_____

_____

Weight Is a Measure of Gravitational Force (p. 128)

17. The strength of the gravitational force exerted by an object

depends on the _____ of the object. The measure of the Earth's gravitational force on an object is the

object's _____.

Identify each of the following statements as describing mass or weight. Write *M* for mass and *W* for weight.

18. _____ different on the moon

19. _____ expressed in newtons

20. _____ measured in grams

21. _____ measure of gravitational force

22. _____ value doesn't change

23. _____ amount of matter in an object

24. On Earth, mass and weight are constant, which means they are

the same thing. True or False? (Circle one.)

Review (p. 129)

Now that you've finished Section 4, review what you learned by answering the Review questions in your ScienceLog.

CHAPTER

6 DIRECTED READING WORKSHEET

Forces in Motion

As you read Chapter 6, which begins on page 136 of your textbook, answer the following questions.

Imagine . . . (p. 136)

1. What is the Vomit Comet?

_____

_____

_____

_____

2. In this chapter you will learn how _____

affects the _____ of objects and how the

_____ of _____

apply to your life.

What Do You Think? (p. 137)

Answer these questions in your ScienceLog now. Then later, you'll have a chance to revise your answers based on what you've learned.

Investigate! (p. 137)

3. What is the purpose of this activity?

_____

_____

_____

Section 1: Gravity and Motion (p. 138)

4. Do you agree with what Aristotle might say, that the basketball would land first, then the baseball, then the marble? Explain.

_____

_____

_____

_____

_____

Chapter 6, continued

_____

All Objects Fall with the Same Acceleration (p. 138)

5. Did Galileo prove Aristotle wrong? Explain.

_____

_____

_____

_____

_____

6. What does 9.8 m/s/s have to do with acceleration?

_____

_____

_____

Air Resistance Slows Down Acceleration (p. 139)

7. Why does a crumpled piece of paper hit the ground before a flat sheet of paper?

_____

_____

8. Air resistance is affected by the _____ and _____ of an object.

9. Air resistance matches the _____ when the net force equals zero. (acceleration or force of gravity)

10. When a falling object stops _____, it has reached _____ velocity.

11. If there were no air resistance, hailstones would
   a. hit the Earth at velocities near 350 m/s.
   b. float gently to the ground like snowflakes.
   c. melt before they hit the ground.
   d. behave exactly as they do now.

12. A sky diver experiences free fall. True or False? (Circle one.)

13. Free fall occurs because of high air resistance. True or False? (Circle one.)

Orbiting Objects Are in Free Fall (p. 141)

14. An astronaut is weightless in space. True or False? (Circle one.)

Chapter 6, continued

15. The shuttle in Figure 7 follows the curve of the Earth's surface as

it moves _____ at a constant speed. At

the same time, it is in _____ because of
the Earth's gravity.

16. Why don't astronauts hit their head on the ceiling of the falling
shuttle?

_____

_____

_____

_____

17. Earth's gravity provides a _____ force
that keeps the moon in orbit.

Projectile Motion and Gravity (p. 143)

18. The projectile motion of a leaping frog has two components—

_____ and _____ .

Mark each of the following statements *True* or *False*.

19. _____ The components of projectile motion affect each
other.

20. _____ Horizontal motion of an object is parallel to the
ground.

21. _____ Ignoring air resistance, the horizontal velocity of a
thrown object never changes.

22. _____ On Earth, gravity gives thrown objects their down-
ward vertical motion.

23. If you shoot an arrow aimed directly at the bull's-eye of your target, where will the arrow hit your target? Why?

_____

_____

_____

Review (p. 144)

Now that you've finished Section 1, review what you've learned by
answering the Review questions in your ScienceLog.

Section 2: Newton's Laws of Motion (p. 145)

1. In 1686, _____ published *Principia,* a

work explaining _____ laws to help
people understand how forces relate to the

_____ of objects.

Newton's First Law of Motion (p. 145)

2. What is Newton's first law?

_____

_____

_____

_____

3. An object in motion would keep moving forever if it never ran

into another object or an unbalanced force. True or False?
(Circle one.)

4. _____ is the unbalanced force that slows
down sliding desks, rolling baseballs, and moving cars.

5. How does inertia explain why it would be so difficult to play
softball with a bowling ball?

_____

_____

_____

_____

_____

_____

Newton's Second Law of Motion (p. 148)

6. What is Newton's second law of motion?

_____

_____

_____

_____

7. Look at the Environmental Science Connection. A small car with a small engine cannot accelerate as well as a large car with a large engine. True or False? (Circle one.)

8. An object's acceleration decreases as the force on it increases. True or False? (Circle one.)

9. Force equals _____ times

_____ .

10. The watermelon in Figure 16 has more _____ and _____ than the apple, so the watermelon is harder to move than the apple.

Review (p. 149)

Now that you've finished the first part of Section 2, review what you've learned by answering the Review questions in your ScienceLog.

Newton's Third Law of Motion (p. 150)

11. What is Newton's third law of motion?

_____

_____

_____

_____

12. The phrase "equal and opposite" means that the action force and the reaction force have the same _____ but act in opposite _____ .

_____

Chapter 6, continued
_____

13. What action and reaction forces are present when you are sitting on a chair?

_____

_____

_____

_____

_____

14. In a force pair, the reaction and action forces affect the same object. True or False? (Circle one.)

15. When a ball falls off a ledge, gravity pulls the ball toward Earth and also pulls Earth toward the ball. True or False? (Circle one.)

Momentum Is a Property of Moving Objects (p. 152)

16. Why does it take longer for a large truck to stop than it does for a compact car to stop, even though they are traveling at the same velocity and the same braking force is applied?

_____

_____

17. Momentum depends on the _____ and

_____ of an object.

18. In Figure 19, during the collision, the momentum of the cue ball
    a. is added to the total momentum.
    b. is transferred to the billiard ball.
    c. is transferred to the table holding the balls up.
    d. stays with the cue ball.

19. The law of conservation of momentum states that any time two or more objects interact, they may exchange momentum, but the total amount of momentum stays the same. True or False? (Circle one.)

Review (p. 153)

Now that you've finished Section 2, review what you've learned by answering the Review questions in your ScienceLog.

CHAPTER

7 ▸ **DIRECTED READING WORKSHEET**

Forces in Fluids

As you read Chapter 7, which begins on page 160 of your textbook, answer the following questions.

Imagine . . . (p. 160)

1. The _____ is the only place on Earth deep enough to swallow the tallest mountain in the world.

2. How might the design of *Deep Flight* allow it to go to such depths and move quickly through the ocean?

   _____

   _____

   _____

   _____

   _____

What Do You Think? (p. 161)

Answer these questions in your ScienceLog now. Then later, you'll have a chance to revise your answers based on what you've learned.

Investigate! (p. 161)

3. How do you think this activity would demonstrate the effect of pressure on a fluid?

   _____

   _____

   _____

Section 1: Fluids and Pressure (p. 162)

4. How are dogs, flies, dolphins, and humans connected by fluids?

   _____

   _____

5. What else can a fluid do besides flow?

   _____

   _____

6. What can particles in a fluid do that particles in a solid can't do?

    a. They can stay rigidly in place.

    b. They can melt.

    c. They can move easily past each other.

    d. None of the above

All Fluids Exert Pressure (p. 162)

7. Why does a tire expand as you pump air into it?

_____

_____

_____

_____

8. To calculate pressure _____ the force
exerted by a fluid by the _____ over
which the force is exerted. (multiply or divide, area or volume)

9. Can you blow a bubble that is cube-shaped? Explain.

_____

_____

_____

_____

_____

Atmospheric Pressure (p. 163)

Choose the item in Column B that best matches each phrase in
Column A, and write the corresponding letter in the space provided.

Column A	Column B
_____ **10.** pressure caused by the weight of the atmosphere	**a.** 10
_____ **11.** percentage of gases found within 10 km of Earth's surface	**b.** atmospheric pressure
_____ **12.** holds the atmosphere in place	**c.** 80
_____ **13.** newtons pressing on every square centimeter of your body	**d.** gravity

---

Chapter 7, continued

---

14. The depth of a fluid is related to the pressure it exerts. The deeper you go in a fluid, the _____ the pressure becomes. (lower or greater)

15. Look at Figure 4 and number the following locations in order from lowest (1) to highest (5) pressure.

_____ Mount Everest's peak

_____ La Paz, Bolivia

_____ Airplane at cruising altitude

_____ Malibu Beach

_____ 150,000 m above sea level

16. Why do your ears pop as atmospheric pressure decreases?

_____

_____

_____

_____

_____

Review (p. 164)

Now that you've finished the first part of Section 1, review what you learned by answering the Review questions in your ScienceLog.

Water Pressure (p. 165)

17. _____ pressure and

_____ pressure are the two kinds of pressure that contribute to the total pressure on an underwater object.

18. Would you feel more pressure 5 m underwater in a pool or 2 m underwater in a lake? Explain.

_____

_____

_____

_____

19. Water is denser than air. True or False? (Circle one.)

Chapter 7, continued

Fluids Flow from High Pressure to Low Pressure (p. 166)

20. How do you use pressure to sip a drink through a straw?

_____

_____

_____

_____

_____

Read each of the following statements, and describe what happens under the conditions described.

21. Pressure is lower inside the lungs than outside the lungs.

_____

22. Pressure is higher inside a tube of toothpaste than outside the tube.

_____

23. Pressure is higher inside a soda can than outside the can.

_____

Pascal's Principle (p. 167)

24. How does a pumping station affect your shower?

_____

_____

_____

_____

25. Liquids transmit pressure _____ efficiently than gases do because liquids compress _____ easily than gases. (more or less, more or less)

26. A hydraulic brake system in a car acts as a force multiplier because the pistons that push the brake pads are

_____ than the piston that is pushed by the brake pedal. (larger or smaller)

Review (p. 167)

Now that you've finished Section 1, review what you learned by answering the Review questions in your ScienceLog.

Section 2: Buoyant Force (p. 168)

1. How does the buoyant force cause the rubber duck to float?

_____

_____

_____

Buoyant Force Is Caused by Differences in Fluid Pressure (p. 168)

2. Buoyant force exists because the pressure in a fluid is greater at

the _____ of an object. (top or bottom)

3. What did Archimedes figure out about the buoyant force of an object?

_____

_____

_____

4. Buoyant force is determined by
   a. the weight of the displaced water.
   b. the volume of the displaced water.
   c. the weight of the object.
   d. the volume of the object.

Weight vs. Buoyant Force (p. 169)

5. If the weight of the water an object displaces is equal to the

weight of the object, the object _____ .
(floats or sinks)

6. If the weight of the water an object displaces is less than the

weight of the object, the object _____ .
(floats or sinks)

Chapter 7, continued

Choose the statement in Column B that best matches the description
in Column A, and write the corresponding letter in the space provided.

Column A	Column B
_____ **7.** A rock sinks to the bottom of a pond.	**a.** buoyant force < weight
_____ **8.** A duck is buoyed to the surface of a pond.	**b.** buoyant force = weight
_____ **9.** A fish floats in a pond.	**c.** buoyant force > weight

An Object Will Float or Sink Based on Its Density (p. 170)

10. How does the density of a rock affect its ability to float?

_____

_____

11. Why don't most substances float in air if they float in water?

_____

_____

12. The helium balloon in Figure 11 floats in air because the dis-

placed air is _____ than the helium.
(lighter or heavier)

The Mystery of Floating Steel (p. 171)

13. How does the shape of a steel ship allow the ship to float?

_____

_____

_____

_____

14. When you flood a submarine's tanks with sea water

    a. the submarine becomes less dense and rises.
    b. the submarine becomes less dense and sinks.
    c. the submarine becomes denser and rises.
    d. the submarine becomes denser and sinks.

Chapter 7, continued

---

15. How does a fish's swim bladder cause the fish to move like a submarine?

_____

_____

_____

_____

_____

Review (p. 172)

Now that you've finished Section 2, review what you learned by answering the Review questions in your ScienceLog.

Section 3: Bernoulli's Principle (p. 173)

1. What happens to your shower curtain when you increase the water pressure in your shower?

_____

_____

Fluid Pressure Decreases as Speed Increases (p. 173)

2. What did Bernoulli say about the speed of a moving fluid?

   a. The faster the speed, the higher the pressure.
   b. The slower the speed, the lower the pressure.
   c. The faster the speed, the lower the pressure.
   d. The slower the speed, the higher the pressure.

3. The table-tennis ball in Figure 14 stays in the water stream. Why?

_____

_____

_____

It's a Bird! It's a Plane! It's Bernoulli's Principle! (p. 174)

4. Look at Figure 15. The shape of an airplane wing causes the air

above the wing to flow _____ than the
air below it. (slower or faster)

5. The upward force acting on an airplane wing due to air flow is

called _____. (buoyancy or lift)

Chapter 7, continued

Thrust and Wing Size Determine Lift (p. 175)

6. How does thrust increase lift?

_____

_____

_____

7. How large must the wings be for each of the following airplane types? (Circle one for each type.)

   a. small commuter plane   small   medium   large

   b. high-performance jet   small   medium   large

   c. glider   small   medium   large

8. Which of the methods below do birds use to stay in the air? (Circle all that apply.)

   a. They use large wing size to glide on wind currents.
   b. They pull up their legs.
   c. They move their tails up and down.
   d. They flap their wings.

Drag Opposes Motion in Fluids (p. 176)

9. In a strong wind, drag is the _____ that you walk against.

10. _____ usually causes drag forces in flight.

11. Wing flaps are parts of a commercial airplane that reduce drag.

   True or False? (Circle one.)

Wings Are Not Always Required (p. 177)

12. Take a moment to examine Figure 19. A baseball with a side spin

   will curve _____ the side where the ball is moving in the same direction as the air flow.
   (away from or toward)

Review (p. 177)

Now that you've finished Section 3, review what you learned by answering the Review questions in your ScienceLog.

CHAPTER

8 **DIRECTED READING WORKSHEET**

Work and Machines

As you read Chapter 8, which begins on page 186 of your textbook, answer the following questions.

Would You Believe . . . ? (p. 186)

1. The Egyptians built the Great Pyramid in 30 years. Why is this so amazing?

_____

_____

_____

2. Which of the following are simple machines used by the Egyptians to build pyramids?

a. screws and screwdrivers
b. inclined planes and levers
c. plows and axes
d. wheels and axles

What Do You Think? (p. 187)

Answer these questions in your ScienceLog now. Then later, you'll have a chance to revise your answers based on what you've learned.

Investigate! (p. 187)

3. What is the purpose of this activity?

_____

_____

_____

Section 1: Work and Power (p. 188)

4. In the scientific sense, you are doing work on this page by reading this question. True or False? (Circle one.)

The Scientific Meaning of *Work* (p. 188)

5. When you bowl, you are applying a _____ to the bowling ball that makes it move down the lane.

6. When you are carrying a heavy suitcase at a constant speed, you are not doing work. Why not?

_____

_____

_____

_____

7. Look at the chart on page 189. Which of the following is having work done on it? (Circle all that apply.)
   a. a grocery bag as you pick it up
   b. a grocery bag as you carry it at a constant speed
   c. a crate as you push it at a constant speed
   d. a backpack as you are walking with a constant speed

Calculating Work (p. 190)

8. Suppose you want to calculate how much work it takes to lift a 160 N barbell. Besides the weight of the barbell, what other information do you need to know?
   a. the shape of the weights
   b. how high the barbell is being lifted
   c. the strength of the person doing the lifting
   d. None of the above

9. In the equation for work, F is the _____

   applied to the object and d is the _____
   through which the force is applied.

Power—How Fast Work Is Done (p. 191)

10. The _____ is the unit used to express power.

11. If you increase power, you are increasing the amount of work done in a given amount of time. True or False? (Circle one.)

Review (p. 191)

Now that you've finished Section 1, review what you learned by answering the Review questions in your ScienceLog.

Section 2: What Is a Machine? (p. 192)

1. The tools used to fix a flat tire are not complicated enough to be

   called machines. True or False? (Circle one.)

Machines—Making Work Easier (p. 192)

2. Which of the following are machines? (Circle all that apply.)

   a. scissors   **c.** screw
   b. lug nut   **d.** car jack

Mark each of the following statements *True* or *False*.

3. _____   As you pry the lid off a can of paint, work is done
                   on the screwdriver and on the lid.

4. _____   While you pry off the lid with the screwdriver, you
                   are not doing any work.

5. _____   The force you apply to the screwdriver while prying
                   is the input force.

6. _____   The work output done by the screwdriver works
                   against the forces of weight and friction to open
                   the lid.

7. _____   The screwdriver helps you open the can because it
                   increases the amount of work you apply to the can.

8. How does using a ramp make lifting a heavy object easier?

   a. The object is moved over a shorter distance.
   b. The ramp increases the amount of work you do.
   c. Less force is needed to move the object over a longer distance.
   d. None of the above

9. When a machine shortens the distance over which a force is

   exerted, the size of the force must _____.
   (increase or decrease)

Mechanical Advantage (p. 196)

10. By comparing the mechanical advantage of two machines, you
    can tell which machine

    a. is bigger.
    b. has a larger output force.
    c. has a larger input force.
    d. makes work easier.

11. Chopsticks allow you to exert force over a longer distance, but the mechanical advantage is

    a. less than one.   **c.** greater than one.

    b. equal to one.   **d.** impossible to determine.

Mechanical Efficiency (p. 197)

12. The work output of a machine is always less than the work input. Where does the missing work go?

    a. It is used to get the machine started.

    b. It is used to overcome the friction created by using the machine.

    c. It is used to keep the machine running.

    d. None of the above

13. Take a moment to read the Life Science Connection in the right column. Your arms can act as machines. The synovial fluid in your elbow joints makes your arms more efficient machines. Why?

_____

_____

14. A machine that has no friction to overcome is called

    a. an ideal machine.   **c.** a joint.

    b. a complex machine.   **d.** a smooth machine.

Review (p. 197)

Now that you've finished Section 2, review what you learned by answering the Review questions in your ScienceLog.

Section 3: Types of Machines (p. 198)

1. Name the six simple machines.

_____

_____

Levers (p. 198)

Mark each of the following statements *True* or *False*.

2. _____ A first-class lever changes the direction of the input force.

3. _____ The output force of a second class lever is smaller than the input force.

4. _____ Third-class levers do not increase the input force.

_____

Chapter 8, continued

_____

Inclined Planes (p. 200)

5. You must divide the _____ of the

inclined plane by the _____ you are
lifting the load in order to calculate mechanical advantage.

6. An inclined plane saves the work required to lift an object.

True or False? (Circle one.)

Wedges (p. 201)

7. Which of the following is NOT a wedge?

    a. knife   **c.** chisel

    b. plow   **d.** ramp

8. How do you calculate the mechanical advantage of a wedge?

_____

_____

Screws (p. 201)

9. Which of the following is NOT a screw?

    a. jar lid   **c.** bolt

    b. steering wheel   **d.** All three are screws.

10. Like using an inclined plane, using a screw enables you to apply

a _____ force over

_____ distance.

Use what you've learned in the first part of Section 3 to answer
the following questions. Choose the machine in Column B that best
matches the definition in Column A, and write the corresponding
letter in the space provided.

Column A	Column B
_____ **11.** a straight, slanted surface	**a.** screw
_____ **12.** a bar that pivots at a fixed point	**b.** inclined plane
_____ **13.** a double inclined plane that moves	**c.** lever
_____ **14.** an inclined plane wrapped in a spiral	**d.** wedge

Review (p. 202)

Now that you've finished the first part of Section 3, review what you
learned by answering the Review questions in your ScienceLog.

▲▲ CHAPTER 8

Wheel and Axle (p. 202)

15. In a wheel and axle, which is larger: the radius of the wheel or the radius of the axle?

_____

16. The input force of a wheel and axle travels along

    a. a circular distance.
    b. a rectangular distance.
    c. an inclined plane.
    d. a spiral.

Pulleys (p. 203)

17. Which of the following is NOT true of pulleys?

    a. A pulley is a grooved wheel that holds a rope or cable.
    b. A movable pulley moves up with the load as it is lifted.
    c. Fixed and movable pulleys working together form a block and tackle.
    d. Fixed pulleys increase force.

Compound Machines (p. 204)

18. A compound machine consists of two or more

_____ .

19. A can opener is a compound machine. Which of the following simple machines are part of a can opener? (Circle all that apply.)

    a. lever   **c.** screw
    b. wheel and axle   **d.** wedge

20. A machine with many moving parts generally has a lower mechanical efficiency than a machine with fewer moving parts.

True or False? (Circle one.)

Review (p. 205)

Now that you've finished Section 3, review what you learned by answering the Review questions in your ScienceLog.

CHAPTER

9 ◼ **DIRECTED READING WORKSHEET**

Energy and Energy Resources

As you read Chapter 9, which begins on page 212 of your textbook, answer the following questions.

Strange but True! (p. 212)

1. What vast treasure-troves have been buried at sea for millions of years?

   a. gold   **c.** salt

   b. gas hydrates   **d.** sodium bicarbonate

2. Scientists suspect that large areas off the coasts of North Carolina

and South Carolina may contain _____

times the natural gas consumed by the United States in 1 year.

3. What happens when you hold a flame near icy formations of water and methane?

_____

_____

What Do You Think? (p. 213)

Answer these questions in your ScienceLog now. Then later, you'll have a chance to revise your answers based on what you've learned.

Investigate! (p. 213)

4. What will you find out in this activity?

_____

_____

Section 1: What Is Energy? (p. 214)

5. Where do you think energy is being transferred as the tennis game is played?

_____

_____

_____

Energy and Work—Working Together (p. 214)

6. Energy is the _____ to do work.

7. When you hit a tennis ball with a racket, energy is transferred from the racket to the ball. True or False? (Circle one.)

8. Work and energy are both measured in

   _____ .

Kinetic Energy Is Energy of Motion (p. 215)

9. Does the tennis player have kinetic energy if she isn't moving? Explain.

   _____

   _____

   _____

   _____

10. In Figure 2, swinging a hammer gives the hammer energy to do work. True or False? (Circle one.)

11. Which of the following have kinetic energy? (Circle all that apply.)
    a. a falling raindrop   c. a plane in the sky
    b. a rolling bowling ball   d. a parked car

12. Which of the following is NOT true of kinetic energy?
    a. The faster something moves, the more kinetic energy it has.
    b. The lower the mass is, the higher the kinetic energy.
    c. Speed has a greater effect on kinetic energy than mass has.

13. The truck and the red car in Figure 3 are traveling at the same speed. So why does the truck have more kinetic energy?

   _____

   _____

Potential Energy Is Energy of Position (p. 216)

14. Why does a stretched bow have potential energy?

   _____

   _____

   _____

Chapter 9, continued

---

15. Take a moment to look at Figure 5. Which of the following would have more gravitational potential energy than a diver on a platform? (Circle all that apply.)

  a. a diver with the same mass on a lower platform
  b. a diver with the same mass on a higher platform
  c. a diver with more mass on the same platform
  d. a diver with less mass on the same platform

16. What two measurements do you multiply together to get gravitational potential energy?

  _____

  _____

Mechanical Energy Sums It All Up (p. 217)

17. The mechanical energy of the juggler's pins in Figure 6 is the

total energy of motion and position of the pins. True or False? (Circle one.)

18. Potential energy plus gravitational energy equals mechanical

energy. True or False? (Circle one.)

Review (p. 217)

Now that you've finished the first part of Section 1, review what you learned by answering the Review questions in your ScienceLog.

Forms of Energy (p. 218)

19. List the six forms of energy.

  _____

  _____

20. The total potential energy of all the particles in an object is

known as thermal energy. True or False? (Circle one.)

21. In Figure 7, the particles in ocean water have less kinetic energy than the particles in steam. Why?

  _____

  _____

  _____

  _____

▲ ▲
▲ ▲ **CHAPTER 9**
▲

Choose the type of energy in Column B that best matches the definition in Column A, and write the corresponding letter in the space provided. The type of energy may be used more than once.

Column A	Column B
____ **22.** energy produced by vibrations of electrically charged particles	**a.** chemical
____ **23.** energy of a compound that changes when its atoms are rearranged to form a new compound	**b.** electrical
____ **24.** energy caused by an object's vibrations	**c.** light
____ **25.** energy of moving electrons	**d.** sound
____ **26.** energy used in radar systems	

27. Nuclear energy can be produced only by splitting the nucleus of an atom. True or False? (Circle one.)

28. Where does the sun get its energy to light and heat the Earth?

_____

_____

_____

29. The nucleus of an atom can store _____ energy. (potential or kinetic)

Review (p. 221)

Now that you've finished Section 1, review what you learned by answering the Review questions in your ScienceLog.

Section 2: Energy Conversions (p. 222)

1. When you are hammering a nail, what is one type of energy conversion that is taking place?

_____

_____

_____

2. An energy conversion can happen between any two forms of energy. True or False? (Circle one.)

_____

Chapter 9, continued

_____

From Kinetic to Potential and Back (p. 222)

Take a look at Figure 14. Mark each of the following energy conversions $K{\rightarrow}P$ (kinetic to potential) or $P{\rightarrow}K$ (potential to kinetic).

3. _____ You jump down, and the trampoline stretches.

4. _____ The trampoline does work on you, and you bounce up.

5. _____ You reach the top of your jump on the trampoline.

6. _____ You are about to hit the trampoline again.

7. In Figure 15, the potential energy of the pendulum is the smallest at which point in its swing?

   a. the highest point   **c.** the slowest point
   b. the lowest point   **d.** none of the above

Conversions Involving Chemical Energy (p. 223)

8. Why does eating breakfast give you energy to start the day?

_____

_____

_____

_____

9. The energy you get from food originally comes from the sun.

True or False? (Circle one.)

10. During _____ , plants convert light energy into chemical energy.

Conversions Involving Electrical Energy (p. 225)

11. Look at Figure 18. Which of the following forms of energy are converted from electrical energy when you turn on a hair dryer? (Circle all that apply.)

   a. nuclear energy   **c.** kinetic energy
   b. sound energy   **d.** thermal energy

12. In a battery, _____ energy is converted

into _____ energy.

Review (p. 225)

Now that you've finished the first part of Section 2, review what you learned by answering the Review questions in your ScienceLog.

Energy and Machines (p. 226)

13. How does a machine make work easier? (Circle all that apply.)

    a. by changing the direction of the required force
    b. by changing the size of the required force
    c. by requiring no force
    d. by increasing the amount of energy transferred

14. A nutcracker can transfer more energy to a nut than you transfer to the nutcracker. True or False? (Circle one.)

15. Which of the following kinetic energy transfers does NOT occur when you ride a bike as in Figure 20?

    a. from legs to pedals   **c.** from chain to back wheel
    b. from pedals to chain   **d.** from gear wheel to chain

16. An energy conversion makes the telephone a useful machine. Explain.

_____

_____

_____

17. As gasoline burns inside an engine, _____ energy is converted into thermal and kinetic energy. (electrical or chemical)

Why Energy Conversions Are Important (p. 228)

18. Take a look at Figure 22. How could wind help you cook a meal?

_____

_____

_____

19. The more efficient the light bulb is, the more electrical energy is converted into light energy instead of thermal energy.

True or False? (Circle one.)

20. Energy output is always more than energy input. True or False? (Circle one.)

Review (p. 228)

Now that you've finished Section 2, review what you learned by answering the Review questions in your ScienceLog.

Section 3: Conservation of Energy (p. 229)

1. A roller-coaster car never returns to its starting height because
   energy gets lost along the way. True or False? (Circle one.)

Where Does the Energy Go? (p. 229)

2. Where does friction oppose motion on a roller-coaster car?
   (Circle all that apply.)
   a. between the wheels of the car and the track
   b. between the car and the surrounding air
   c. between the car's axles and the wheels
   d. between the car and the passenger

3. The original amount of potential energy of a roller-coaster car is

   converted into kinetic energy and _____
   energy as the car races down the hill. (electrical or thermal)

4. The potential energy of the car at the top of the second hill of a
   roller coaster is equal to the original potential energy of the car

   at the top of the first hill. True or False? (Circle one.)

Energy Is Conserved Within a Closed System (p. 230)

5. A roller coaster is involved in a closed system. List its parts.

   _____

6. The law of conservation of energy states that energy can be neither

   _____ nor _____ .

7. Look at Figure 24. Which forms of energy are converted from the
   electrical energy that enters the light bulb? (Circle all that apply.)
   a. light energy
   b. thermal energy warming the bulb
   c. thermal energy caused by friction in the wire
   d. chemical energy

No Conversion Without Thermal Energy (p. 231)

8. During an energy conversion, energy is rarely converted to

   thermal energy. True or False? (Circle one.)

9. Why is it impossible to make a perpetual motion machine?

_____

_____

_____

_____

_____

_____

_____

Review (p. 231)

Now that you've finished Section 3, review what you learned by answering the Review questions in your ScienceLog.

Section 4: Energy Resources (p. 232)

1. How do we use energy resources?

_____

_____

_____

_____

2. The _____ is the energy resource responsible for most other energy resources.

Nonrenewable Resources (p. 232)

3. Which of the following are fossil fuels? (Circle all that apply.)

   a. coal   **c.** petroleum
   b. wood   **d.** natural gas

4. Fossil fuels are formed from the remains of plants and animals

that lived millions of years ago. True or False? (Circle one.)

5. Explain why fossil fuels are concentrated forms of the sun's energy.

_____

_____

_____

_____

_____

_____

Use the images on page 233 to answer questions 6–8.

6. Most of the coal supply in the United States is used for heating.

True or False? (Circle one.)

7. Which of the following is NOT a petroleum product?

a. rayon clothing   **c.** petrochemicals
b. a candle   **d.** wool

8. The cleanest-burning fossil fuel is _____ .

9. Take a moment to study Figure 27. Put the following events in the correct sequence for the production of electrical energy from fossil fuels by writing the appropriate number in the space provided.

_____ A large magnet spins within a ring of wire coils.

_____ Thermal energy converts liquid water to steam.

_____ Electric current is generated in the wire coils.

_____ Fossil fuels are burned.

_____ Steam pushes against the blades of a turbine.

_____ Electrical energy is distributed to homes.

_____ Water is pumped into a boiler.

10. In nuclear fission, a nucleus of a(n) _____ element releases energy when it is split into two nuclei.

11. Nuclear energy is considered to be a renewable resource.

True or False? (Circle one.)

12. One uranium fuel pellet in Figure 28 contains the energy equiva-lent of about one _____ of coal.

Renewable Resources (p. 235)

13. Solar energy can be used to run a television. Explain.

_____

_____

_____

_____

14. Electrical energy generated from falling water is called hydro-gravity. True or False? (Circle one.)

15. A wind turbine converts the _____ energy of the wind into electrical energy. (potential or kinetic)

16. Geothermal energy results from the heating of Earth's

_____ .

17. Which of the following is NOT an example of biomass?

    a. leaves   **c.** wood

    b. steel   **d.** cactus

18. Corn can be used to make a cleaner-burning fuel for cars.

True or False? (Circle one.)

The Two Sides to Energy Resources (p. 237)

Choose the energy resource in Column B that best matches the disadvantage in Column A, and write the corresponding letter in the space provided.

Column A	Column B
_____ **19.** requires large areas of farmland	**a.** fossil fuels
_____ **20.** produces radioactive waste	**b.** nuclear
_____ **21.** requires dams that disrupt river ecosystems	**c.** solar
_____ **22.** expensive for large-scale energy production	**d.** water
_____ **23.** waste water can damage soil	**e.** wind
_____ **24.** only practical in windy areas	**f.** geothermal
_____ **25.** burning produces smog and acid rain	**g.** biomass

Review (p. 237)

Now that you've finished Section 4, review what you learned by answering the Review questions in your ScienceLog.

CHAPTER

10 DIRECTED READING WORKSHEET

The Energy of Waves

As you read Chapter 10, which begins on page 244 of your textbook, answer the following questions.

This Really Happened! (p. 244)

1. How did a land-based earthquake cause the worst marine disaster in the history of the town of Kodiak?

_____

_____

_____

_____

_____

_____

What Do You Think? (p. 245)

Answer these questions in your ScienceLog now. Then later, you'll have a chance to revise your answers based on what you've learned.

Investigate! (p. 245)

2. What is the purpose of this activity?

_____

_____

_____

Section 1: The Nature of Waves (p. 246)

3. Which of the following waves might your family have experienced after a day at the beach? (Circle all that apply.)

   a. water waves

   b. microwaves

   c. light waves

   d. sound waves

Waves Carry Energy (p. 246)

4. A wave can carry energy away from its source. True or False? (Circle one.)

Chapter 10, continued

5. Air doesn't travel with sound waves. Give an example of what would happen if it did.

   _____

   _____

   _____

6. Take a moment to examine Figure 1. Floating birds and boats bob up and down and travel in the same direction as waves.

   True or False? (Circle one.)

7. Waves can transfer _____ through the

   vibration of _____ in solid, liquid, or gaseous media.

8. If you put an alarm clock inside a jar and remove all the air from the jar, you wouldn't hear the clock ringing. Why?

   _____

   _____

   _____

   _____

   _____

9. Electromagnetic waves are waves that require a medium to

   transfer energy. True or False? (Circle one.)

10. Take a moment to look at the Brain Food in the left column of page 248. How did Percy Spencer accidentally discover microwave cooking?

    _____

    _____

    _____

    _____

11. Which of the following lists of waves contains ONLY mechanical waves?

    a. radio waves, X rays, sound waves
    b. seismic waves, water waves, microwaves
    c. microwaves, water waves, X rays
    d. sound waves, seismic waves, water waves

Types of Waves (p. 249)

After reading pages 249–251 in your text, match each type of wave in Column B to the correct statements in Column A, and write the corresponding letter in the appropriate space. Wave types can be used more than once.

Column A	Column B
____ **12.** Particles move forward at the crest and backward at the trough.	**a.** longitudinal wave
____ **13.** Particles vibrate up and down.	**b.** transverse wave
____ **14.** Electromagnetic waves are this type of wave.	**c.** surface wave
____ **15.** A rarefaction is a section of this type of wave that is less crowded than normal.	
____ **16.** Particles vibrate back and forth along the path the wave travels.	
____ **17.** A sound wave is this type of wave.	
____ **18.** Two types of waves can combine to form this type of wave.	
____ **19.** Particles travel perpendicular to the direction the wave travels.	
____ **20.** This wave occurs at or near the boundary of two media.	

21. Take a moment to read the Earth Science Connection on the right side of page 251. List the following types of seismic waves in order from least to greatest destructive power: surface, longitudinal, transverse.

_____

Review (p. 251)

Now that you've finished Section 1, review what you learned by answering the Review questions in your ScienceLog.

Chapter 10, continued

Section 2: Properties of Waves (p. 252)

1. Waves made by the breeze were very different than waves created by the speedboat. Describe the difference.

_____

_____

_____

_____

Amplitude (p. 252)

2. Which of the following is NOT true of amplitude?

a. The smaller the amplitude, the more energy that is carried by a wave.

b. It is the maximum distance a wave vibrates from its resting position.

c. The larger the amplitude of a water wave, the taller the wave.

d. None of the above are true.

Wavelength (p. 253)

3. Wavelength can be measured between corresponding points on

two adjacent waves. True or False? (Circle one.)

Frequency (p. 254)

4. The frequency of a wave is the _____ of waves produced in a given amount of time. To measure frequency of a transverse wave, I can count the number of

_____ or _____ that pass a point in a certain amount of time. I can also count

the number of _____ or

_____ for a longitudinal wave.

Mark each of the following statements *True* or *False*.

5. _____ The frequency of a wave is not related to its wavelength.

6. _____ When wavelength decreases, frequency increases.

7. _____ A wave with a long wavelength carries less energy than a wave with a short wavelength.

8. _____ Frequency is expressed in hertz (Hz).

Wave Speed (p. 255)

9. The speed of a wave is the _____ traveled

by a wave in a given amount of _____ .

Wave speed is equal to _____ times

_____ .

10. Complete the MathBreak in your ScienceLog. Then answer the
following questions:

a. What can you calculate if you know the frequency and speed
of a wave?

_____

_____

b. If you know the speed of a wave, and you want to find its
wavelength, what other piece of information would be helpful
for you to know?

_____

_____

_____

Review (p. 255)

Now that you've finished Section 2, review what you learned by
answering the Review questions in your ScienceLog.

Section 3: Wave Interactions (p. 256)

1. The moon doesn't produce light like the stars do. So why does the
moon shine?

_____

_____

Waves Bounce Back During Reflection (p. 256)

2. Reflection is when waves hit a barrier and some of them pass

through it. True or False? (Circle one.)

3. Water waves cannot be reflected. True or False? (Circle one.)

4. Sound waves that reflect off canyon walls or classroom walls are

called _____ .

Changing Speed Bends Waves During Refraction (p. 257)

5. Why does the pencil in a half-filled glass of water look like it's broken?

_____

_____

_____

_____

6. A wave bends when it enters a new medium because the part of the wave that enters first is traveling at a different speed than the rest of the wave. True or False? (Circle one.)

Waves Bend Around Barriers or Through Openings During Diffraction (p. 257)

7. Which of the following are true of diffraction? (Circle all that apply.)

   a. Waves curve or bend when they reach the edge of an object.
   b. Waves bend through an opening.
   c. Sound travels around buildings.
   d. The size of the barrier does not affect the amount of diffraction.

8. Diffraction of a wave depends on what two things?

_____

_____

_____

9. Look at Figure 18 on page 258. If the opening is larger than the wavelength, _____ diffraction occurs. (a little or a lot of)

Overlapping Waves Cause Interference (p. 258)

10. Which of the following is TRUE about overlapping waves?

   a. They share the same space.
   b. They cannot be in the same place at the same time.
   c. They pass around each other.

11. Interference is the result of two or more waves combining to form three waves. True or False? (Circle one.)

Match each type of interference in Column B to the correct statements in Column A, and write the corresponding letter in the appropriate space. Interference types can be used more than once.

Column A	Column B
_____ **12.** The crests of two waves overlap.	**a.** constructive interference
_____ **13.** The resulting wave has a smaller amplitude than the original waves had.	**b.** destructive interference
_____ **14.** Waves of the same amplitude cancel each other out.	
_____ **15.** The result is a wave with deeper troughs and higher crests than the original waves.	
_____ **16.** Crests of one wave overlap with the troughs of another wave.	

17. A standing wave occurs from interference between the original wave and the reflected wave. True or False? (Circle one.)

18. In a standing wave, _____ interference causes portions of the wave to have a large amplitude, while total _____ interference causes other portions of the wave to be at rest.

19. Standing waves only *look* like they're standing still. What's really going on?

  a. Waves are vibrating faster than light.
  b. Waves are no longer vibrating.
  c. Waves are traveling in both directions.

20. Place the following steps in the correct sequence to explain how the marimba player in Figure 23 uses resonance. Write the appropriate number in the space provided.

  _____ The amplitude of the vibrations is increased.

  _____ The bar vibrates.

  _____ The marimba player strikes a bar.

  _____ A loud note is produced.

  _____ The air in the column underneath the bar absorbs energy from the vibrating bar and begins to vibrate.

  _____ The frequency of the air column matches the frequency of the bar.

  _____ The air column resonates with the bar.

21. During resonance, a vibrating object will cause a second object to vibrate when it reaches the second object's resonant frequency.

True or False? (Circle one.)

22. How did the Tacoma Narrows Bridge earn the nickname Galloping Gertie?

   a. The bridge was used most often by people traveling on horseback.
   b. The bridge experienced wavelike motions during strong winds.
   c. The bridge was built by Gertrude Stein in July 1940.
   d. Soldiers marched across it in such a way that it collapsed in 1831.

23. How did resonance contribute to the destruction of Galloping Gertie?

_____

_____

_____

_____

_____

_____

_____

_____

_____

_____

_____

Review (p. 261)

Now that you've finished Section 3, review what you learned by answering the Review questions in your ScienceLog.

CHAPTER

11 **DIRECTED READING WORKSHEET**

Introduction to Electricity

As you read Chapter 11, which begins on page 268 of your textbook, answer the following questions.

Strange but True! (p. 268)

1. Which of the following is NOT true about electric eels?

   a. An electric eel uses electric discharges to stun or kill its prey.
   b. The body of an adult eel can generate 5,000 to 6,000 V.
   c. The eel's thick skin protects it from electrocuting itself.
   d. Eels swallow their prey whole.

2. The bursts of voltage that an eel gives off when it shocks its prey

   is greater than the voltage of an electrical outlet. True or False? (Circle one.)

What Do You Think? (p. 269)

Answer these questions in your ScienceLog now. Then later, you'll have a chance to revise your answers based on what you've learned.

Investigate (p. 269)

3. What is the purpose of this activity?

   _____

   _____

   _____

Section 1: Electric Charge and Static Electricity (p. 270)

4. When you shuffle your feet on the carpet on a dry day, you get a shock from the metal objects that you touch. What is the cause of this?

   _____

   _____

   _____

   _____

Chapter 11, continued

Atoms and Charge (p. 270)

Choose the word in Column B that best matches the description in Column A, and write the corresponding letter in the space provided.

Column A	Column B
_____ **5.** composed of atoms	**a.** proton
_____ **6.** a positively charged particle of the nucleus	**b.** neutron
_____ **7.** a negatively charged particle found outside the nucleus	**c.** matter
_____ **8.** a particle of the nucleus that has no charge	**d.** electron

9. According to the law of electric charges, like charges attract and opposite charges repel. True or False? (Circle one.)

10. Why don't electrons fly out of atoms while traveling around the nucleus?

_____

_____

11. Which of the following does not determine the strength of an electric force between charged objects?
   a. the age of the charges
   b. the size of the charges
   c. the distance between the charges
   d. All of the above contribute.

12. The region around a charged particle that can exert a force on another charged particle is called the _____.

Charge It! (p. 272)

Choose the word in Column B that best matches the definition in Column A, and write the corresponding letter in the space provided.

Column A	Column B
_____ **13.** "wiping" electrons off of one object onto another	**a.** induction
_____ **14.** transfer of electrons when one object touches another object	**b.** friction
_____ **15.** rearranging electrons in an uncharged object when it is near a charged object	**c.** conduction

16. When objects are charged, charges cannot be created or destroyed. True or False? (Circle one.)

17. An electroscope can determine which of the following?
   a. whether or not an object is charged
   b. the material that a charged object is made of
   c. the strength of the charge on an object
   d. how many electrons are involved in the charge

Review (p. 274)

Now that you've finished the first part of Section 1, review what you learned by answering the Review questions in your ScienceLog.

Moving Charges (p. 274)

18. Electric cords are often covered in plastic and have metal prongs.

This is because metal is a good _____ and

plastic is a good _____ .

19. Hair dryers should not be used near water. Why?

_____

_____

_____

_____

Static Electricity (p. 275)

20. What is static electricity?
   a. an electric charge on a stationary object
   b. random electric signals from your dryer
   c. the buildup of electric charges on an object
   d. electricity that moves away from an object

21. How does clothing that has become charged in a dryer lose its charge?

_____

_____

22. As charges move off an object, the object loses its static electricity. This process is called electric _____

23. An electric discharge can occur quickly or slowly. True or False? (Circle one.)

24. Standing on a beach or golf course during a thunderstorm can make you like a lightning rod. Why?

_____

_____

_____

_____

_____

Review (p. 277)

Now that you've finished Section 1, review what you learned by answering the Review questions in your ScienceLog.

Section 2: Electrical Energy (p. 278)

1. Name something that uses electrical energy that would be difficult for you to live without. Explain.

_____

_____

_____

Batteries Are Included (p. 278)

Choose the word in Column B that best matches the definition in Column A, and write the corresponding letter in the space provided.

Column A	Column B
_____ **2.** converts chemical energy into electrical energy	**a.** electrode
_____ **3.** type of cell that contains a solid or pastelike electrolyte	**b.** cell
	c. electrolyte
_____ **4.** mixture of chemicals in a cell	**d.** wet
_____ **5.** where charges enter or exit a cell	**e.** dry
_____ **6.** type of cell that contains a liquid electrolyte	

Bring on the Potential (p. 279)

7. In a battery, electric current exists between the two electrodes

because there is a difference in _____
between the electrodes.

Chapter 11, continued

8. If the potential difference is increased, the current

_____ . (increases or decreases)

Other Ways of Producing Electrical Energy (p. 280)

Choose the term in Column B that best matches the description in
Column A, and write the corresponding letter in the space provided.

Column A	Column B
_____ **9.** converts kinetic energy into electrical energy	**a.** silicon
_____ **10.** converts light energy into electrical energy	**b.** copper
_____ **11.** ejects electrons when struck by light	**c.** thermocouple
_____ **12.** converts thermal energy into electrical energy	**d.** solar panel
_____ **13.** one type of wire used in some thermocouples	**e.** generator

Review (p. 280)

Now that you've finished Section 2, review what you learned by
answering the Review questions in your ScienceLog.

Section 3: Electric Current (p. 281)

1. Where does most of the electrical energy used in your home come
from?

   a. large rechargeable batteries
   b. electric power plants
   c. chemical reactions
   d. small generators

Current Revisited (p. 281)

2. In a wire, electrons travel at almost the speed of light.

   True or False? (Circle one.)

Label each of the following as a characteristic of alternating current
or direct current. Write *AC* for alternating current or *DC* for direct
current.

3. _____ It's produced by batteries.

4. _____ It provides the energy in your home.

5. _____ The flow of charges switches directions.

6. _____ Charges flow in one direction.

7. _____ It's more practical for transferring electrical energy.

Voltage (p. 282)

8. Another word for potential difference is _____

which is expressed in _____ .

9. As voltage increases in a circuit, the current decreases.

  True or False? (Circle one.)

10. The electrical outlets in the United States usually supply a

voltage of _____ V.

11. Look at the Life Science Connection on page 283. Why do doctors intentionally create a large voltage across the chest of a heart-attack victim?

  _____

  _____

Resistance (p. 283)

12. Which of the following associations is false?

  a. Increasing resistance increases current.
  b. Decreasing resistance increases current.
  c. Good conductors have low resistance.
  d. Poor conductors have high resistance.

13. An object's resistance depends on which of the following properties of the object? (Circle all that apply.)

  a. thickness
  b. length
  c. temperature
  d. color

14. The light bulb in Figure 18 has a filament made of tungsten. Why is tungsten used in this bulb?

  _____

  _____

  _____

15. Thin wires have _____ resistance than

thick wires. Short wires have _____ resistance than long wires.

16. The resistance of metals generally increases with increasing temperatures because at higher temperatures, the faster-moving atoms slow down the flow of electric charge.

  True or False? (Circle one.)

Ohm's Law: Putting It All Together (p. 285)

17. In the equation for Ohm's law, what do the letters *I, V,* and *R* stand for?

_____

_____

18. Who was Georg Ohm?

    a. an electrician   **c.** an inventor

    b. a professor   **d.** an author

19. If you know the current produced in a wire and the voltage applied, you can calculate the resistance of the wire.

    True or False? (Circle one.)

Electric Power (p. 285)

20. Electric power is expressed in

    a. ohms.   **c.** amperes.

    b. volts.   **d.** watts.

21. Light bulbs may be labeled "100 W" or "40 W." This describes

    a. how long they burn.

    b. how they are disposed of.

    c. how fast the light travels.

    d. how bright they glow.

22. A television uses more power than a hair dryer.

    True or False? (Circle one.)

Measuring Electrical Energy (p. 286)

23. In the equation for electrical energy, what do *E, P,* and *t* stand for?

_____

_____

24. What do electric meters measure?

    a. power   **c.** current

    b. voltage   **d.** energy

Review (p. 287)

Now that you've finished Section 3, review what you learned by answering the Review questions in your ScienceLog.

Section 4: Electric Circuits (p. 288)

1. An electric circuit always begins and ends in the same place.

    True or False? (Circle one.)

Parts of a Circuit (p. 288)

2. Which of the following are parts of all electric circuits?
(Circle all that apply.)

  a. a load   **c.** wires
  b. an energy source   **d.** a switch

3. A _____ opens and closes a circuit.

Types of Circuits (p. 289)

4. A _____ circuit has all parts connected in
a single loop. A _____ circuit has different loads on separate branches.

5. A string of holiday lights wired together in series has a burned-
out bulb. Why do all of the lights go out?

_____

_____

_____

6. If a break occurs in one of the loops of a parallel circuit, current
will not run through the other loops. True or False? (Circle one.)

Household Circuits (p. 292)

7. Which of the following may cause a circuit failure?
(Circle all that apply.)

  a. water   **c.** too many loads
  b. broken wires   **d.** excess insulation

8. As more loads are added to a parallel circuit, the entire circuit
draws more current. True or False? (Circle one.)

9. How does a fuse disrupt the flow of charges when the current is
too high?

  a. A metal strip warms up and bends away from the circuit wires.
  b. A metal strip warms up and melts, leaving a gap.
  c. A metal strip changes from a conductor to an insulator.
  d. None of the above

10. Circuit breakers are inconvenient because breakers must be
replaced when they are tripped. True or False? (Circle one.)

Review (p. 293)

Now that you've finished Section 4, review what you learned by
answering the Review questions in your ScienceLog.

CHAPTER

12 **DIRECTED READING WORKSHEET**

Introduction to Atoms

As you read Chapter 12, which begins on page 302 of your textbook, answer the following questions.

Would You Believe . . . ? (p. 302)

1. What do dinosaurs have in common with atoms?

_____

_____

2. How did scientists find information that caused them to change their theory about the way *T. rex* walked? (Circle all that apply.)

a. by studying well-preserved dinosaur tracks

b. by examining similarities between the skeletons of *T. rex* and an ostrich

c. by observing a *T. rex* as it was walking

d. by extracting DNA from fossilized mosquitoes

3. Scientists are able to develop theories about dinosaurs and

atoms only through _____ evidence. (direct or indirect)

What Do You Think? (p. 303)

Answer these questions in your ScienceLog now. Then later, you'll have a chance to revise your answers based on what you've learned.

Investigate! (p. 303)

4. How do you think rolling marbles in this activity will help you identify the mystery object?

_____

_____

_____

_____

_____

Section 1: Development of the Atomic Theory (p. 304)

5. Atoms are NOT

    a. a relatively new idea to us.

    b. the building blocks of all matter.

    c. the smallest particles into which an element can be divided and still be the same substance.

    d. seen with the scanning tunneling microscope.

6. An explanation that is supported by testing and brings together a broad range of hypotheses and observations is called a

_____ .

Democritus Proposes the Atom (p. 304)

7. The word *atom* comes from a Greek word that means

_____ . (invisible or indivisible)

8. Which of the following statements is part of Democritus's theory about atoms?

    a. Atoms are small, soft particles.

    b. Atoms are always standing still.

    c. Atoms join together to form different materials.

    d. None of the above

9. We know that Democritus was right about atoms. So why did people ignore Democritus's ideas for such a long time?

_____

_____

_____

Dalton Creates an Atomic Theory Based on Experiments (p. 305)

10. By conducting experiments and making observations, Dalton figured out that elements combine in random proportions because they're made of individual atoms. True or False? (Circle one.)

11. Dalton's theory states that atoms cannot be

_____ , _____ , or

_____ .

12. Atoms of different elements are exactly alike.

True or False? (Circle one.)

---

13. How did Dalton think atoms formed new substances?

_____

_____

Thomson Finds Electrons in the Atom (p. 306)

Mark the following statements *True* or *False*.

14. _____ In 1897, J. J. Thomson made a discovery that proved the first part of Dalton's atomic theory was correct.

15. _____ Thomson discovered that there were small particles inside the atom.

16. _____ Thomson found that the electrically charged plates affected the direction of a cathode-ray tube beam.

17. _____ Thomson knew the beam was made of particles with a positive charge because it was pulled toward a positive charge.

18. When you rub a balloon on your hair, your hair is

_____ to the balloon because both the

hair and the balloon have become _____.

19. The two types of charges are positive and neutral.

True or False? (Circle one.)

20. Objects with the same charges attract each other. True or False? (Circle one.)

21. In Thomson's "plum-pudding" model, electrons are NOT

    a. negatively charged.
    b. present in every type of atom.
    c. collected together in the center of the atom.
    d. scattered throughout a blob of positively charged material.

Review (p. 307)

Now that you've finished the first part of Section 1, review what you learned by answering the Review questions in your ScienceLog.

Rutherford Opens an Atomic "Shooting Gallery" (p. 308)

22. Before his experiment, Rutherford expected the particles to

deflect to the sides of the gold foil. True or False? (Circle one.)

23. Figure 6 shows the new atomic model resulting from Rutherford's experiment. Which of the following statements is NOT part of Rutherford's revision of his former teacher's atomic theory?

a. Atoms are made mostly of empty space.
b. The nucleus is a dense, charged center of the atom.
c. Lightweight, negative electrons move in the nucleus.
d. Most of an atom's mass is in the nucleus.

24. The diameter of the nucleus of an atom is

_____ times smaller than the diameter of the atom.

Bohr States That Electrons Can Jump Between Levels (p. 310)

25. In Bohr's atomic model, _____ travel

in definite paths around the _____ at

specific levels. Each level is a certain _____

from the nucleus. Electrons cannot travel between levels, but they can

_____ from level to level.

26. Bohr's model only predicted some atomic behavior.

True or False? (Circle one.)

The Modern Theory: Electron Clouds Surround the Nucleus (p. 310)

27. The exact path of a moving electron can now be predicted.

True or False? (Circle one.)

28. What are electron clouds?

_____

_____

_____

Review (p. 310)

Now that you've finished Section 1, review what you learned by answering the Review questions in your ScienceLog.

Section 2: The Atom (p. 311)

1. In this section you will learn about the

_____ that act on the particles inside the

atom.

How Small Is an Atom? (p. 311)

Each of the following statements is false. Change the underlined
word to make the statement true. Write the new word in the space
provided.

2. A sheet of aluminum foil is about <u>500</u> atoms thick.

_____

3. <u>An Olympic medal</u> contains about twenty thousand billion

billion atoms of copper and zinc. _____

What's Inside an Atom? (p. 312)

Choose the term in Column B that best matches the phrase in
Column A, and write the appropriate letter in the space provided.

Column A	Column B
____ **4.** particle found in the nucleus that has no charge	**a.** electron cloud
____ **5.** particle found in the nucleus that is positively charged	**b.** electron
____ **6.** atom with an unequal number of protons and electrons	**c.** amu
____ **7.** negatively charged particle found outside the nucleus	**d.** nucleus
____ **8.** size of this determines the size of the atom	**e.** proton
____ **9.** contains most of the mass of an atom	**f.** ion
____ **10.** SI unit used for the masses of atomic particles	**g.** neutron

Review (p. 313)

Now that you've finished the first part of Section 2, review what you
learned by answering the Review questions in your ScienceLog.

How Do Atoms of Different Elements Differ? (p. 313)

11. The simplest atom is the _____ atom. It
has one proton and one electron.

12. Neutrons in the atom's nucleus keep the protons from moving

apart. True or False? (Circle one.)

13. To build an atom of _____ , you need two
protons, two neutrons, and two electrons.

14. Each element is composed of atoms with the same number of

_____ . (neutrons or protons)

Are All Atoms of an Element the Same? (p. 314)

15. It is NOT true that isotopes of an element

   a. have the same number of protons but different numbers of
neutrons.

   b. are stable when radioactive.

   c. share most of the same chemical properties.

   d. share most of the same physical properties.

How Do You Calculate the Mass of an Element? (p. 316)

16. The weighted average of the masses of all the naturally occurring

isotopes of an element is called _____ mass.

What Forces Are at Work in Atoms? (p. 317)

Choose the type of force in Column B that best matches the phrase in
Column A, and write the corresponding letter in the space provided.

Column A	Column B
____ **17.** counteracts the electromagnetic force so protons stay together in the nucleus	**a.** gravity
____ **18.** depends on the mass of objects and the distance between them	**b.** electromagnetic force
____ **19.** plays a key role in neutrons changing into protons in unstable atoms	**c.** strong force
____ **20.** holds the electrons around the nucleus	**d.** weak force

Review (p. 317)

Now that you've finished Section 2, review what you learned by
answering the Review questions in your ScienceLog.

CHAPTER

13 **DIRECTED READING WORKSHEET**

The Periodic Table

As you read Chapter 13, which begins on page 324 of your textbook, answer the following questions.

Would You Believe . . . ? (p. 324)

1. Hyraxes are related to elephants, even though they don't look alike. What have scientists similarly discovered about different-looking elements?

_____

_____

_____

2. The periodic table is useful for _____

known elements and predicting the _____
of unknown elements.

What Do You Think? (p. 325)

Answer these questions in your ScienceLog now. Then later, you'll have a chance to revise your answers based on what you've learned.

Investigate! (p. 325)

3. What will you be looking for in this activity?

_____

_____

_____

Section 1: Arranging the Elements (p. 326)

4. Why do you think scientists might have been frustrated by the organization of the elements before 1869?

_____

_____

_____

_____

_____

Chapter 13, continued

Discovering a Pattern (p. 326)

5. Mendeleev spent a lot of train rides organizing the elements according to their properties. Which arrangement of elements produced a repeating pattern of properties?

   a. by increasing density

   b. by increasing melting point

   c. by increasing shine

   d. by increasing atomic mass

6. How are the days of the week periodic?

_____

_____

_____

_____

7. Mendeleev noticed after arranging the elements that similar

_____ and _____ properties could be observed in every

_____ element.

8. Mendeleev was able to predict the properties of elements that no one knew about. How was this possible?

_____

_____

_____

_____

Changing the Arrangement (p. 327)

9. A few elements in Mendeleev's table seemed to be mysteriously out of place according to their properties. How did Moseley solve the mystery? (Circle all that apply.)

   a. He rearranged the elements by atomic number.

   b. He discovered protons, neutrons, and electrons.

   c. He disproved the periodic law.

   d. He determined the number of protons in an atom.

10. The basis of the periodic table is the periodic

_____, which states that the properties of

elements are _____ of their atomic

_____.

Use the periodic table on pages 328–329 of your text to fill in the answers to the following questions.

11. Which information is NOT included in each square of the periodic table in your text?

    a. atomic number   **c.** melting point

    b. chemical symbol   **d.** atomic mass

12. How can you tell that chlorine is a gas at room temperature?

_____

_____

_____

_____

_____

13. Rows of elements are called _____ , and

columns of elements are called _____ or

_____ .

14. Who will approve the names of the newest elements?

    a. the scientist who discovered each element

    b. an international committee of scientists

    c. the chemists from a research institute

15. Silicon is a _____ .
    (metal, nonmetal, or metalloid)

Finding Your Way Around the Periodic Table (p. 330)

16. The _____ of elements determine
whether they are classified as metals, nonmetals, or metalloids.
They can also be placed in categories according to the number of

_____ in the outer

_____ level of an atom.

17. There is a zigzag line on the periodic table. How can it help you?

_____

_____

_____

CHAPTER 13 ▲▲▲

Chapter 13, continued

Use the pictures on pages 330–331 to help you match the category in Column B with the description in Column A, and write the corresponding letter in the space provided. Categories may be used more than once.

Column A	Column B
____ **18.** few electrons in the outer energy level	**a.** metals
____ **19.** have some properties of the other two categories	**b.** nonmetals
____ **20.** brittle and nonmalleable solids	**c.** metalloids
____ **21.** complete or almost-complete set of electrons in the outer energy level	
____ **22.** conducts heat from a stovetop to your food	
____ **23.** can prevent a spark from igniting gasoline in your car	
____ **24.** half-complete shell of electrons in the outer energy level	
____ **25.** formed into electrical wires	
____ **26.** flattened into sheets of food wrap without shattering	
____ **27.** border the zigzag line on the periodic table	

28. Some elements are named after scientists, like Einstein, and places, like California. True or False? (Circle one.)

29. The chemical symbol Pb comes from the

_____ word *plumbum*, which means

_____ .

30. What happens as you move from left to right through each period on the periodic table?

a. Elements change from having properties of nonmetals to having properties of metals.

b. Elements change from having properties of metalloids to having properties of metals.

c. Elements change from liquids to gases.

d. None of the above

Review (p. 333)

Now that you've finished Section 1, review what you learned by answering the Review questions in your ScienceLog.

CHAPTER
14 **DIRECTED READING WORKSHEET**

Chemical Bonding

As you read Chapter 14, which begins on page 350 of your textbook, answer the following questions.

Strange but True! (p. 350)

1. A scientist discovered superglue by accident. What was he trying to develop?

   _____

   _____

   _____

2. What type of bond is the force that holds atoms together?

   a. electrical   **c.** gravitational
   b. chemical   **d.** static

3. Which of the following are possible uses of superglue?
   (Circle all that apply.)

   a. attaching aircraft parts
   b. repairing a cracked tooth
   c. fertilizing plants

What Do You Think? (p. 351)

Answer these questions in your ScienceLog now. Then later, you'll have a chance to revise your answers based on what you've learned.

Investigate! (p. 351)

4. What will you be observing in this activity?

   _____

   _____

   _____

Section 1: Electrons and Chemical Bonding (p. 352)

5. Every substance in the world can be made out of about 100

   elements. True or False? (Circle one.)

Atoms Combine Through Chemical Bonding (p. 352)

6. Sugar is made from atoms of which of the following elements? (Circle all that apply.)

    a. carbon   **c.** hydrogen

    b. nitrogen   **d.** oxygen

7. A chemical bond is the _____ of attraction that holds a pair of atoms together.

Electron Number and Organization (p. 352)

8. In order to make the overall charge of an atom zero, there must be an

equal number of negatively charged _____

and positively charged _____ .

9. Valence electrons are the electrons in an atom's innermost

energy level. True or False? (Circle one.)

Look at Figure 3 on page 354. Write the number of valence electrons for each of the following elements:

10. _____ oxygen

11. _____ sodium

12. _____ chlorine

13. _____ helium

To Bond or Not to Bond (p. 354)

14. Which electrons determine whether or not an atom will form bonds?

    a. the electrons in the nucleus

    b. the electrons in the innermost energy level

    c. the electrons in the outermost energy level

    d. None of the above

15. An atom will not normally form a chemical bond if it has

_____ valence electrons.

16. Which of the following does NOT describe how atoms can fill their outermost energy level?

    a. by sharing electrons with other atoms

    b. by losing electrons to other atoms

    c. by gaining electrons from other atoms

    d. by gaining kinetic energy from other atoms

Chapter 14, continued

---

17. Why is a helium atom stable with only two electrons in its outer-most energy level?

_____

_____

_____

Review (p. 355)

Now that you've finished Section 1, review what you learned by answering the Review questions in your ScienceLog.

Section 2: Types of Chemical Bonds (p. 356)

1. Three kinds of chemical bonds are _____,

_____ , and _____ .

Ionic Bonds (p. 356)

2. Describe how two atoms can become ions.

_____

_____

_____

3. An atom that loses one or more electrons from its outermost

energy level becomes a positively charged ion. True or False?
(Circle one.)

4. Which of the following elements give up electrons to other
atoms? (Circle all that apply.)

   a. sodium   **c.** chlorine
   b. aluminum   **d.** oxygen

5. Why do the elements in Groups 1 and 2 react so easily?

_____

_____

_____

6. Atoms of nonmetals lose one or more protons when they form

ionic bonds. True or False? (Circle one.)

▲▲ CHAPTER 14
▲

7. The names of negative ions that form when atoms gain electrons

   have the ending _____ .

8. A large amount of energy is released when atoms of Group 17

   elements lose electrons. True or False? (Circle one.)

9. Which of the following are common properties of an ionic
   compound? (Circle all that apply.)

   a. Its solid form is a crystal lattice.
   b. It contains alternating positive and negative ions.
   c. It is soft and pliable at room temperature.
   d. Its positive and negative ions repel each other.
   e. It has a low melting point.
   f. It has a high boiling point.
   g. It is neutral.

10. Look at Figure 8 on page 359. What force causes both the forma-
    tion of ionic bonds and static cling?

    a. the Earth's gravity
    b. the repulsion of like charges
    c. the attraction of opposite charges
    d. a magnetic pole

Review (p. 359)

Now that you've finished the first part of Section 2, review what you
learned by answering the Review questions in your ScienceLog.

Covalent Bonds (p. 360)

11. Covalent bonds form between atoms that require a large amount

    of energy in order to lose an electron. True or False? (Circle one.)

12. In a covalent bond, neither atom loses or gains an electron.

    Instead, one or more electrons are _____
    by the atoms. (shared or created)

13. Look at Figure 11. The electrons that are shared by two atoms
    spend most of their time

    a. near the smallest of the two atoms.
    b. near the largest of the two atoms.
    c. between the nuclei of the two atoms.
    d. in the nuclei of the two atoms.

14. A group of atoms held together by covalent bonds forms a

    neutral particle called a _____ .

15. Draw the electron-dot diagram for krypton.

16. Draw the electron-dot diagram for water.

17. In an electron-dot diagram, each dot represents one proton.
True or False? (Circle one.)

18. Diatomic molecules are the simplest kinds of molecules. They
consist of two atoms bonded together. True or False?
(Circle one.)

19. Give three examples of complex molecules.

_____

_____

20. Carbon is known as the building block of life. Which of the
following is a property of this important element?
(Circle all that apply.)

   a. Each of its atoms needs to make four bonds.
   b. It is found in all proteins.
   c. It can bond with other elements and form long chains.
   d. It is in a water molecule.

Metallic Bonds (p. 363)

21. In a metal, a sea of "swimming" protons surround the metallic
ions. True or False? (Circle one.)

22. What properties of metals result from metallic bonding?

_____

_____

_____

_____

_____

▲ ▲ ▲ **CHAPTER 14**

23. Because ions in a metal can be easily rearranged without break-ing the metallic bonds, metals tend to be easily

    a. shattered.   **c.** reshaped.

    b. crystallized.   **d.** broken.

24. Which of the following is NOT a typical property of a metal?

    a. malleability   **c.** conductivity

    b. ductility   **d.** brittleness

25. Besides being valuable in the jewelry industry, gold is special because it can be hammered into a very thin foil. This property is called

    a. malleability.   **c.** conductivity.

    b. ductility.   **d.** brittleness.

Identify each of the following substances as containing mostly ionic, mostly covalent, or mostly metallic bonds. Refer back to the earlier parts of Section 2 as needed. Write *I* for ionic, *C* for covalent, and *M* for metallic.

26. _____ copper wire

27. _____ water

28. _____ table salt

29. _____ sugar

30. _____ carbon dioxide

31. _____ plaster of Paris

32. _____ aluminum foil

33. _____ gold jewelry

34. _____ sea shells

Review (p. 365)

Now that you've finished Section 2, review what you learned by answering the Review questions in your ScienceLog.

CHAPTER

15 DIRECTED READING WORKSHEET

Chemical Reactions

As you read Chapter 15, which begins on page 372 of your textbook, answer the following questions.

Imagine . . . (p. 372)

The moment an airbag-equipped vehicle slams into a wall, a sequence of events rapidly takes place to protect you from hitting the dashboard. Place the events in the proper sequence by writing the appropriate number in the space provided.

1. _____ A small electric current is sent to the gas generator.

2. _____ The airbag fills the space between you and the dashboard.

3. _____ A sensor detects the sudden decrease in speed.

4. _____ The airbag inflates with gas formed in the gas generator.

5. _____ Chemicals in the gas generator react, creating a gas.

6. Could you construct an airbag using vinegar and baking soda for the gas generator? Explain.

_____

_____

_____

_____

_____

What Do You Think? (p. 373)

Answer these questions in your ScienceLog now. Then later, you'll have a chance to revise your answers based on what you've learned.

Investigate! (p. 373)

7. What is the purpose of this activity?

_____

_____

_____

Section 1: Forming New Substances (p. 374)

8. The color of leaves that contain chlorophyll is

_____ .

9. The red, orange, and yellow colors of leaves are always present

but are hidden until the chlorophyll breaks down. True or False? (Circle one.)

Chemical Reactions (p. 374)

10. Chemical reactions do not change the substances involved.

True or False? (Circle one.)

11. Look at Figure 2. What chemical reaction causes bubbles to form in a muffin?

_____

_____

_____

_____

Choose the clue to a chemical reaction in Column B that best matches the example in Column A, and write the corresponding letter in the space provided.

Column A	Column B
_____ **12.** heat produced by a fire	**a.** color change
_____ **13.** precipitate	**b.** energy change
_____ **14.** bubbles	**c.** solid formation
_____ **15.** white spots caused by bleach	**d.** gas formation

16. What does a chemical reaction have to do with making and breaking chemical bonds?

_____

_____

_____

_____

_____

Chapter 15, continued

Chemical Formulas (p. 376)

17. The subscript in the chemical formula H_2O tells you there are two

    a. atoms of hydrogen in the molecule.
    b. electrons on the hydrogen atom in the molecule.
    c. elements in the molecule.
    d. atoms of oxygen in the molecule.

In the space provided, write the number of atoms of each element in each of the following chemical formulas.

18. O_2 _____

19. $C_6H_{12}O_6$ _____

20. H_2O _____

21. $Ca(NO_3)_2$ _____

22. Covalent compounds are usually composed of two or more

    _____ .

23. In the chart on page 377, what number does the prefix *hepta-* stand for?

    a. six   **c.** seven
    b. eight   **d.** five

Write the formula for each of the following covalent compounds.

24. dinitrogen monoxide _____

25. carbon dioxide _____

26. Ionic compounds are composed of a _____

    and a _____ .

27. The overall charge of an ionic compound is zero. True or False? (Circle one.)

Write the formula for each of the following ionic compounds.

28. sodium chloride _____

29. magnesium chloride _____

Chemical Equations (p. 378)

30. What do musical notation and chemical equations have in common?

_____

_____

_____

CHAPTER 15

31. When carbon reacts with oxygen to form carbon dioxide,

carbon dioxide is the _____ .
(product or reactant)

32. Look at Figure 9. In a chemical equation, the formulas of the

_____ appear before the arrow, and

the formulas of the _____ appear after
the arrow.

33. Which of the following are diatomic elements?
(Circle all that apply.)

   a. oxygen   **d.** bromine
   b. carbon   **e.** nitrogen
   c. neon   **f.** hydrogen

34. If you make a mistake in a chemical formula, you may be

describing a very different substance. True or False? (Circle one.)

35. According to the Brain Food on page 379, what is one good
reason to use hydrogen gas as a fuel?

_____

_____

_____

36. The coefficient 2 in $2CO_2$ means that there are

   a. two oxygen atoms and one carbon atom present.
   b. two oxygen atoms present.
   c. two carbon atoms and two oxygen atoms present.
   d. two carbon dioxide molecules present.

37. After you finish looking at Figure 11 on page 380, balance the
following chemical equation: $O_2 + H_2 \rightarrow H_2O$.

_____

38. Chemical equations must be balanced. Why?

_____

_____

_____

_____

_____

_____

39. Antoine Lavoisier's work led to the law of

_____ of mass, which states that

mass is neither _____ nor

_____ in chemical or physical changes.

Review (p. 381)

Now that you've finished Section 1, review what you learned by answering the Review questions in your ScienceLog.

Section 2: Types of Chemical Reactions (p. 382)

1. Which of the following is NOT one of the four classifications of chemical reactions discussed in the text?

    a. synthesis   **c.** single-replacement

    b. decomposition   **d.** double-decomposition

Synthesis Reactions (p. 382)

2. In a synthesis reaction, a single compound is formed from two or more substances. True or False? (Circle one.)

Decomposition Reactions (p. 383)

3. Decomposition reactions are the _____ of synthesis reactions.

Single-Replacement Reactions (p. 383)

4. How is a person who cuts in on a dancing couple like a single-replacement reaction?

_____

_____

_____

_____

_____

5. In a single-replacement reaction, a more reactive element can replace a less reactive element from a compound.

True or False? (Circle one.)

Double-Replacement Reactions (p. 384)

6. _____ in two _____
switch places in a double-replacement reaction.

After reading Section 2, match these reaction types with their correct examples. Choose the type of reaction from Column B that best matches the example in Column A, and write the corresponding letter in the space provided.

Column A	Column B
____ **7.** Zinc reacts with hydrochloric acid to form zinc chloride and hydrogen.	**a.** decomposition
____ **8.** Water can be broken down to form hydrogen and oxygen.	**b.** double-replacement
____ **9.** Magnesium reacts with oxygen in the air to form magnesium oxide.	**c.** single-replacement
____ **10.** Sodium nitrate and silver chloride are formed from the reaction of sodium chloride with silver nitrate.	**d.** synthesis

Review (p. 384)

Now that you've finished Section 2, review what you learned by answering the Review questions in your ScienceLog.

Section 3: Energy and Rates of Chemical Reactions (p. 385)

1. The rate at which a chemical reaction occurs cannot be changed.

True or False? (Circle one.)

Every Reaction Involves Energy (p. 385)

2. Chemical bonds break as they _____
energy. (absorb or release)

3. An exothermic reaction releases energy. True or False? (Circle one.)

4. Look at Figure 20. Give one example of how energy is released during a chemical reaction.

_____

_____

Chapter 15, continued

---

5. In an endothermic reaction, the chemical energy of the

   _____ is less than the chemical energy of

   the _____ .

6. What is the law of conservation of energy?

   _____

   _____

   _____

7. What happens to the energy absorbed by endothermic reactions?

   _____

   _____

8. Which of the following statements about chemical reactions is
   NOT true?
   a. Exothermic reactions release energy.
   b. Energy can be stored in the bonds of a molecule.
   c. The activation energy of a chemical reaction is the amount of
      energy released.
   d. All chemical reactions require some energy to get started.

9. Once an exothermic reaction is started, it continues to supply
   the activation energy necessary for breaking the bonds in the

   reactant particles. True or False? (Circle one.)

Factors Affecting Rates of Reactions (p. 387)

10. What are four factors that affect how rapidly a chemical
    reaction takes place?

    _____

    _____

    _____

    _____

11. The light stick in Figure 23 glows brighter in hot water because

    the rate of a reaction _____ as tempera-
    ture increases. (decreases or increases)

12. The concentration of a solution is a measure of the amount of

    one substance dissolved in another. True or False? (Circle one.)

13. How does increasing concentration increase the rate of reaction? (Circle all that apply.)

    a. There are more reactant particles present, so the particles are more likely to collide with each other.

    b. The distance between particles is smaller, so the particles are able to collide more frequently.

    c. More particles present gives a lowered activation energy.

    d. Increasing the concentration of a solution lowers the surface area of the reactants.

14. What is one way you can increase the surface area of a solid reactant?

_____

_____

_____

15. A catalyst lowers the activation energy of a reaction.

    True or False? (Circle one.)

16. The catalysts used in your body are called

    _____ .

17. Which of the following slows down a reaction?

    a. platinum in a catalytic converter

    b. a food preservative

    c. an enzyme in your body

    d. None of the above

18. Which of the following can increase the rate of a chemical reaction? (Circle all that apply.)

    a. raising the temperature

    b. reducing the amount of one substance dissolved in another

    c. increasing the surface area of a solid reactant

    d. adding a catalyst

    e. adding an inhibitor

Review (p. 389)

Now that you've finished Section 3, review what you learned by answering the Review questions in your ScienceLog.

CHAPTER

Chemical Compounds

As you read Chapter 16, which begins on page 396 of your textbook, answer the following questions:

Strange but True . . . (p. 396)

1. What was the inventor of Silly Putty® trying to do?

_____

_____

_____

What Do You Think? (p. 397)

Answer these questions in your ScienceLog now. Then later, you'll have a chance to revise your answers based on what you've learned.

Investigate! (p. 397)

2. The purpose of this activity is to see how the type of

_____ in compounds can determine

the _____ of the compounds.

Section 1: Ionic and Covalent Compounds (p. 398)

3. Which of the following is NOT true about chemical compounds?

    a. Chemical compounds are composed of molecules or ions.
    b. There are actually only a few kinds of chemical compounds.
    c. Chemical compounds have different kinds of bonds.
    d. Chemical compounds are all around us.

Ionic Compounds (p. 398)

4. In an ionic bond, electrons are transferred from metal atoms to

    nonmetal atoms. True or False? (Circle one.)

5. When sodium reacts with chlorine, you get

    _____ .

6. The crystal-lattice structure contains ions arranged in a

    _____ three-dimensional pattern.

7. In a crystal lattice, each ion is surrounded by and bonded to ions

    with the same charge. True or False? (Circle one.)

8. Circle the properties of ionic compounds.

strong bonds   weak bonds

hard to break   shatter easily

high melting point   low melting point

solid at room temperature   liquid at room temperature

difficult to dissolve in water   easy to dissolve in water

9. _____ ionic compounds cannot conduct an electric current. (Dissolved or Undissolved)

Covalent Compounds (p. 399)

10. Which of the following statements is NOT true of covalent compounds?

   a. They form when two atoms share electrons.
   b. They have weaker bonds than ions in a crystal lattice.
   c. They are independent particles called molecules.
   d. They have a higher melting point than ionic compounds.

11. Why don't water and oil mix?

_____

_____

_____

_____

12. Most solutions that contain molecules of covalent compounds do not conduct an electric current. True or False? (Circle one.)

Review (p. 400)

Now that you've finished Section 1, review what you learned by answering the Review questions in your ScienceLog.

Section 2: Acids, Bases, and Salts (p. 401)

1. Lemon changes the color of tea because the lemon contains a substance called a(n) _____ . (acid or base)

Chapter 16, continued

Acids (p. 401)

2. Why should you never use taste as a test to identify an unknown chemical?

_____

_____

3. Acids react with all metals to produce hydrogen gas.

   True or False? (Circle one.)

4. When an acid is placed in water, the number of hydrogen (H^+)

   ions _____ . These extra hydrogen ions

   combine with _____ molecules to form

   _____ ions.

5. A substance that changes color in the presence of an acid or a

   base is called an indicator. True or False? (Circle one.)

6. Blue litmus paper tests for the presence of bases. True or False?
   (Circle one.)

Choose the acid in Column B that best matches each use in Column A, and write the corresponding letter in the space provided. Acids may be used more than once.

Column A	Column B
____ **7.** paper and paint production	**a.** citric acid
____ **8.** the "bite" in soft drinks	**b.** nitric acid
____ **9.** digestion	**c.** sulfuric acid
____ **10.** in orange juice	**d.** hydrochloric acid
____ **11.** algae preventative	**e.** carbonic acid
____ **12.** rubber production	

13. An acid is strong if all of its molecules break apart in water to

    produce hydrogen ions. True or False? (Circle one.)

14. Which of the following are weak acids? (Circle all that apply.)

    sulfuric acid   carbonic acid

    phosphoric acid   citric acid

    nitric acid   hydrochloric acid

Bases (p. 403)

15. Bases have a _____ taste and feel

_____ .

16. Bases increase the number of hydrogen ions in a solution.

True or False? (Circle one.)

Match each of the bases in Column B with the common uses in
Column A, and write the corresponding letter in the space provided.

Column A	Column B
____ **17.** treating heartburn	**a.** ammonia
____ **18.** unclogging drains and making soap	**b.** calcium hydroxide
____ **19.** making cement	**c.** sodium hydroxide
____ **20.** household cleaning	**d.** magnesium hydroxide

21. A base is strong when all the molecules break apart in water to

produce hydrogen ions. True or False? (Circle one.)

Acids and Bases Neutralize One Another (p. 404)

22. How do antacids get rid of heartburn?

_____

23. H^+ ions of an acid and OH^- ions of a base combine to form the

compound we know as _____ .

24. A solution with a pH of 3 is _____ .

(basic or acidic)

25. Which of the following can be used to determine the pH of a
solution? (Circle all that apply.)

    a. a mixture of indicators   **c.** litmus paper
    b. a pH meter   **d.** lemon juice

26. Living things need a steady _____ in
their environment.

27. Why is the helicopter in Figure 15 dumping a base in the lake?

_____

_____

_____

Salts (p. 406)

28. The _____ ion of a base and the

_____ ion of an acid can combine to

form a(n) _____ compound called a salt.

29. Salts can be produced by three types of reactions. List the three reactions shown in Figure 16.

_____

_____

_____

Mark each of the following statements *True* or *False*.

30. _____ Calcium chloride is used to season your food.

31. _____ Salt can keep the roads ice-free in winter.

32. _____ The walls of your room may contain calcium sulfate.

33. _____ Sodium nitrate is used in food preservation.

Review (p. 406)

Now that you've finished Section 2, review what you learned by answering the Review questions in your ScienceLog.

Section 3: Organic Compounds (p. 407)

1. Organic compounds are composed of _____ molecules.

Each Carbon Atom Forms Four Bonds (p. 407)

2. Carbon atoms are able to bond with up to four other atoms.

True or False? (Circle one.)

3. Take a look at Figure 18. Which of the following is NOT a type of carbon backbone?

a. carbon atoms connected in a line
b. a chain continuing in more than one direction
c. a chain forming a ring
d. None of the above

Biochemicals: The Compounds of Life (p. 408)

4. Biochemicals are organic compounds made by living things.

True or False? (Circle one.)

5. What are carbohydrates? (Circle all that apply.)

   a. biochemicals

   b. sources of energy

   c. one or more simple sugars bonded together

Identify each item below as belonging to simple (S) or complex (C) carbohydrates.

6. _____ many sugar molecules bonded together

7. _____ produced by plants through photosynthesis

8. _____ bread, cereal, and pasta

9. _____ stored, extra sugar

10. _____ one or a few sugar molecules bonded together

11. _____ fruits and honey

12. Which of the following is NOT an example of a lipid?

   a. candle   **c.** sugar

   b. chicken fat   **d.** corn oil

13. Lipids store energy, make up cell membranes, and do not dissolve in water. True or False? (Circle one.)

14. Lipids store excess _____ in the body.

Plants tend to store lipids as _____ .

Animals tend to store lipids as _____ .

Mark each of the following statements about phospholipid molecules *True* or *False*.

15. _____ The molecules help control the movement of chemicals into or out of the cell.

16. _____ The head of a molecule is a long carbon backbone composed of only carbon and hydrogen atoms.

17. _____ Water is attracted to the heads of the molecules.

18. _____ The cell membrane is mostly made of three layers of the molecules.

19. Take a look at the Brain Food in the right column of page 409. Do you think we should eliminate cholesterol from our diets? Explain.

_____

_____

_____

20. Biochemicals composed of _____ are called proteins.

21. Which of the following are functions of proteins?
(Circle all that apply.)

    a. regulate chemical activities   **c.** transport and store materials

    b. store excess sugar   **d.** provide support

22. Proteins are all the same size and shape. True or False?
(Circle one.)

23. The shape adopted by the bonded amino acids determines the

function of the protein. True or False? (Circle one.)

24. Insulin is one of the smallest proteins. What very important job
does this small protein have?

_____

_____

25. The protein in your blood that carries oxygen to all parts of your

body is called _____ .

26. Some large proteins help control the transport of materials into

and out of cells. True or False? (Circle one.)

27. Why are nucleic acids sometimes called the "blueprints of life?"

_____

_____

_____

28. Nucleic acids are composed of five building blocks: carbon,

phosphorous, hydrogen, _____ , and

_____ .

Mark each of the following statements *True* or *False*.

29. _____ The two types of nucleic acids are DNA and RNA.

30. _____ RNA is the only genetic material in the cell.

31. _____ DNA molecules can only store a tiny amount of
information due to their small size.

32. _____ RNA contains the information to make DNA.

33. _____ RNA is involved in protein-building.

Review (p. 411)

Now that you've finished the first part of Section 3, review what you
learned by answering the Review questions in your ScienceLog.

Hydrocarbons (p. 412)

For each of the following statements, identify the hydrocarbons as saturated (S), unsaturated (U), or aromatic (A).

34. _____ At least two carbon atoms share a double or a triple bond.

35. _____ Each carbon atom shares a single bond with four other atoms.

36. _____ Air fresheners and mothballs contain this type of hydrocarbon.

37. _____ No atoms can be added without replacing an atom that is part of the hydrocarbon.

38. _____ Most are based on benzene.

39. _____ Many medicines are manufactured using this type of hydrocarbon.

40. _____ Ethene helps ripen fruit.

41. _____ Propane is this type of hydrocarbon.

Other Organic Compounds (p. 413)

42. All organic compounds are made of only carbon and hydrogen.

True or False? (Circle one.)

Match each organic compound in Column B with one of its uses in Column A, and write the corresponding letter in the space provided.

Column A	Column B
_____ **43.** food preservative	**a.** ester
_____ **44.** fragrance	**b.** alkyl halide
_____ **45.** antifreeze	**c.** organic acid
_____ **46.** refrigerant	**d.** alcohol

Review (p. 413)

Now that you've finished Section 3, review what you learned by answering the Review questions in your ScienceLog.

Formation of the Solar System

As you read Chapter 17, which begins on page 422 of your textbook, answer the following questions.

Imagine . . . (p. 422)

1. What kind of gravity would you experience on a space station? Explain.

_____

_____

_____

_____

2. Where could you practice "flying" in a space station?

   a. along the perimeter   **c.** on the floor of the station
   b. at the central axis   **d.** None of the above

What Do You Think? (p. 423)

Answer these questions in your ScienceLog now. Then later, you'll have a chance to revise your answers based on what you've learned.

Investigate! (p. 423)

3. What will you be observing in this activity?

_____

_____

_____

Section 1: A Solar System Is Born (p. 424)

4. The Earth, _____ other planets, and the

   _____ make up our "cosmic neighborhood."

The Solar Nebula (p. 424)

Mark each of the following statements *True* or *False*.

5. _____ The "stuff between the stars" is mostly nothingness.

6. _____ Nebulas are huge interstellar clouds consisting mostly of dust, helium, and hydrogen.

7. _____ Nebulas are the first ingredient for building a new planetary system.

8. _____ In a dense nebula, the strong attraction between the dust and gas particles can pull the mass to the center of the nebula.

9. _____ An increase in temperature results when the gas molecules in a nebula move faster.

10. In Figure 2, how do gravity and pressure keep the gas molecules in balance in a nebula?

_____

_____

_____

_____

_____

11. Which event does NOT cause the formation of a solar nebula?
   a. Something disturbs the balance between pressure and gravity.
   b. Two nebulas crash.
   c. Gas molecules crash between two stars.
   d. A nearby star explodes.

From Planetesimals to Planets (p. 426)

12. Place the following statements in the correct sequence for the formation of a solar system. Figure 3 may help you.

   _____ Dust and clouds of a nebula collapse.

   _____ Planetesimals sweep up dust and gas to form planets.

   _____ Remaining dust and gas are removed from the solar system.

   _____ The nebula flattens into a disk and warms at its center.

   _____ Dust sticks together and forms planetesimals.

13. How did the gas planets become so large?

_____

_____

_____

_____

_____

14. The temperature of the sun caused nearby planets to be made of

a. rocky material.　　　**c.** dust and gases.

b. gases.　　　　　　　**d.** None of the above

15. Where can we find evidence of violent collisions of planetesimals?

_____

16. The center of the solar nebula reached amazing temperatures of

up to _____ million degrees Celsius,

causing hydrogen _____ .

Review (p. 428)

Now that you've finished the first part of Section 1, review what you learned by answering the Review questions in your ScienceLog.

Planetary Motion (p. 429)

17. The planets in the solar system move according to strict physical

laws. True or False? (Circle one.)

Choose the definition in Column B that best matches the term in Column A, and write the corresponding letter in the space provided.

Column A	Column B
_____ **18.** orbit	**a.** time it takes for a body to travel once through its path
_____ **19.** rotation	**b.** spinning on an axis
_____ **20.** period of revolution	**c.** motion of a smaller body in its path around a larger body
_____ **21.** revolution	**d.** path of a body traveling around a larger body

22. The Earth rotates around the sun. True or False? (Circle one.)

23. Why do you suppose the planets don't go flying off into space?

_____

_____

_____

_____

_____

24. What did Kepler observe about the movement of Mars?
  a. It has a circular orbit.
  b. Its moons have different orbits.
  c. It had an ellipse-shaped orbit.
  d. None of the above

Mark each of the following statements *True* or *False*.

25. _____ One astronomical unit (AU) equals 150 million kilometers.

26. _____ Distances from the Earth to other planets can be given in AUs rather than kilometers.

27. _____ Planets move faster when they're far from the sun, and they move slower when they're close to the sun.

28. _____ The time it takes for a planet to travel around the sun can be used to calculate the planet's distance from the sun.

Newton's Law of Universal Gravitation (p. 431)

29. What question was Kepler unable to answer?

_____

_____

30. Newton was able to explain how gravity works. True or False? (Circle one.)

31. Newton's universal law of gravitation tells us that the effect

of _____ on an object depends on the distance from another object and the

_____ of each object.

32. Moving two objects away from each other

_____ the gravitational attraction

between them. (increases or decreases)

33. Because of its velocity and the pull of gravity, the moon stays in

orbit around the Earth. True or False? (Circle one.)

Review (p. 432)

Now that you've finished Section 1, review what you learned by
answering the Review questions in your ScienceLog.

Section 2: The Sun: Our Very Own Star (p. 433)

1. How is our sun like the other stars in our galaxy?

_____

_____

_____

The Structure of the Sun (p. 433)

2. The sun has a solid surface. True or False? (Circle one.)

3. List the layers of the sun in order from innermost to outermost.

_____

_____

Match each of the terms in Column B with the correct description in
Column A, and write the corresponding letter in the space provided.
Figure 9 may help you.

Column A	Column B
_____ **4.** where the sun's energy is produced	**a.** core
_____ **5.** the sun's outer atmosphere	**b.** radiative zone
_____ **6.** where atoms are very closely packed	**c.** convective zone
_____ **7.** where hot and cool gases meet	**d.** photosphere
_____ **8.** only visible during a solar eclipse; deep red	**e.** chromosphere
_____ **9.** the layer of the sun we see	**f.** corona
_____ **10.** can extend outward 10–12 times the sun's diameter	

Chapter 17, continued

Energy Production in the Sun (p. 434)

11. Which of the following are incorrect explanations for the source of the sun's energy? (Circle all that apply.)

    a. The sun is burning fuel to create the heat.
    b. The sun is shrinking.
    c. The sun is bright and hot because of nuclear energy.

12. What did Einstein discover about matter and energy?

    _____

    _____

13. Einstein's formula states that energy equals

    a. mass times the speed of light.
    b. mass times the square of the particles of light.
    c. mass times the square of the speed of light.

14. The source of the sun's energy is _____.

Mark each of the following statements *True* or *False*.

15. _____ In the extreme temperatures of the sun, the repulsive force in atoms is stronger than the attractive force.

16. _____ In nuclear fission, helium turns into hydrogen.

17. _____ The energy released in nuclear fusion is immediately converted into light that leaves the sun.

18. _____ Deuterium is a heavy form of hydrogen.

Activity on the Sun's Surface (p. 436)

19. Sunspots are not really "spots" on the sun. What are they?

    _____

    _____

    _____

20. What happened to the Earth's climate during the years when no sunspots were observed?

    a. The Earth experienced a little ice age.
    b. The Earth's climate stayed the same.
    c. The Earth experienced global warming.

21. Solar flares are giant _____ on the sun's surface.

_____

Chapter 17, continued

_____

22. What are auroras?

_____

_____

_____

_____

Review (p. 437)

Now that you've finished Section 2, review what you learned by
answering the Review questions in your ScienceLog.

Section 3: The Earth Takes Shape (p. 438)

1. How is studying the Earth's early history like trying to put
together a huge jigsaw puzzle?

_____

_____

_____

The Solid Earth Takes Form (p. 438)

Fill in the empty boxes in the table below to complete the explanation of how the Earth was formed.

Cause	Effect
2. Planetesimals accumulated.	
3.	The planet became spherical.
4.	The Earth was made warmer.
5. The inside rock could not cool off as quickly as the temperature increased.	

Chapter 17, continued

6. As the rocks of the young Earth melted, the denser, heavier

materials _____ , while the lighter

materials _____ the surface.

7. The sources of energy for heating the Earth were

_____ materials and heat added by

_____ and other falling materials.

Each of the following phrases refers to a layer of the Earth shown in Figure 17. In the space provided, identify each phrase as describing the Earth's *crust, mantle,* or *core.*

8. _____ a thin skin over the entire planet

9. _____ contains the heaviest materials, such as iron and nickel

10. _____ the middle layer

11. _____ extends to the center of the Earth

12. _____ inhabited by humans

The Atmosphere Evolves (p. 440)

13. Take a look at the Chemistry Connection in the left column. By studying the chemistry of Titan, scientists hope to

    a. figure out how Titan can support human life.
    b. understand the formation of nitrogen.
    c. learn how molecules essential to life may be formed.

14. Over 4 billion years ago, no life existed on Earth. What else was different about Earth at that time?

    _____

    _____

15. Fifty years ago, scientists thought Earth's early atmosphere was composed of mostly methane, ammonia, water, and hydrogen compounds. True or False? (Circle one.)

16. Most of Earth's matter probably came from material similar to

    _____ , while the remaining portion came

    from planetesimals of ice, called _____ .

17. Heated minerals released gases into the Earth's first atmosphere. True or False? (Circle one.)

Chapter 17, continued

18. Which two gases made up the Earth's first atmosphere?

_____

19. Where do scientists think the water for Earth's oceans came from?

   a. comets   c. water vapor
   b. meteoroids   d. other planets

Mark each of the following statements *True* or *False*.

20. _____ The gases in the Earth's second atmosphere came only from volcanoes.

21. _____ The ozone layer was much thicker in Earth's early atmosphere.

22. _____ UV light helped form Earth's current atmosphere.

23. Why is UV light dangerous to your skin?

_____

_____

_____

24. _____ produced oxygen through the process of photosynthesis. These organisms were protected from ultraviolet radiation by a layer of _____.

25. Which of the following events completely changed the Earth's atmosphere?

   a. Comets hit the Earth's surface.
   b. Ultraviolet rays hit bodies of water.
   c. Early life-forms produced oxygen.

Oceans and Continents (p. 443)

26. When did a giant global ocean cover the planet?

_____

27. Continents have existed since the formation of the Earth.

   True or False? (Circle one.)

28. Crust material is heavier than mantle material. True or False? (Circle one.)

29. What does the composition of granite tell geologists about rock in the Earth's crust?

   a. The composition of granite has not changed.
   b. Rocks in the Earth's crust have melted and cooled many times.
   c. The Earth's crust doesn't contain granite.

30. The continents haven't stayed in the same place since their

formation. True or False? (Circle one.)

31. What percentage of the continents had formed by about 3.5 billion years ago?

   a. 100%   **c.** less than 10%
   b. 50%   **d.** 10%

Review (p. 443)

Now that you've finished Section 3, review what you learned by answering the Review questions in your ScienceLog.

CHAPTER

18 DIRECTED READING WORKSHEET

A Family of Planets

As you read Chapter 18, which begins on page 450 of your textbook, answer the following questions.

Imagine . . . (p. 450)

1. Where does the word *planet* come from, and what is its meaning?

_____

_____

2. Which is the largest planet?
   a. Jupiter   **c.** Saturn
   b. Earth   **d.** Neptune

What Do You Think? (p. 451)

Answer these questions in your ScienceLog now. Then later, you'll have a chance to revise your answers based on what you've learned.

Investigate! (p. 451)

3. What is the purpose of this activity?

_____

_____

_____

Section 1: The Nine Planets (p. 452)

4. In 1957, the former Soviet Union launched Sputnik. What was Sputnik?

_____

Measuring Interplanetary Distances (p. 452)

5. The amount of time it takes light to travel around the Earth seven and a half times is _____ .

6. How many light-minutes are there in one astronomical unit?

_____

7. Distances within the solar system must be measured in light-years. True or False? (Circle one.)

The Inner Planets (p. 453)

8. In general, how are the inner planets different from the outer planets? (Circle all that apply.)

    a. They are not spherical.
    b. They are different in size.
    c. They are made of different materials.
    d. They are closer together than the outer planets are.

9. Which group of planets does Earth belong to?

_____

Complete the following section after reading pages 453–457. Each of the statements refers to an inner planet. In the space provided, write *ME* for Mercury, *V* for Venus, or *MA* for Mars.

10. _____ This is the only planet besides Earth with some form of water.

11. _____ On this planet, a day is longer than a year.

12. _____ This planet has the biggest range in surface temperatures.

13. _____ The largest mountain in the solar system is on this planet.

14. _____ On this planet, the sun rises in the west and sets in the east.

15. _____ This planet has the densest atmosphere of all the inner planets.

16. _____ This planet has the thinnest atmosphere.

17. What is the goal of the Earth Science Enterprise? (Circle all that apply.)

    a. to study human effects on the global environment
    b. to study whether life is possible on Mars
    c. to get away from the Earth
    d. to study the Earth using satellites

Review (p. 457)

Now that you've finished the first part of Section 1, review what you learned by answering the Review questions in your ScienceLog.

The Outer Planets (p. 458)

18. Why are the outer planets called gas giants?

_____

_____

Complete the following section after reading pages 458–462. Choose
the planet in Column B that matches the description in Column A,
and write the corresponding letter in the space provided. Planets
may be used more than once.

Column A	Column B
____ **19.** Like Jupiter, this planet is composed of hydrogen and helium.	**a.** Jupiter
____ **20.** This planet is covered by nitrogen ice.	**b.** Saturn
____ **21.** Its moon is more than half its size.	**c.** Uranus
____ **22.** This planet may have been tipped over by a massive object.	**d.** Neptune
____ **23.** The Great Red Spot is on this planet.	**e.** Pluto
____ **24.** This planet is about 15 times more massive than Earth.	
____ **25.** This planet's atmosphere consists of belts of clouds.	
____ **26.** This planet gives off more heat than it receives from the sun.	

27. Which three planets were not known to ancient people?

   a. Saturn, Jupiter, and Pluto   **c.** Jupiter, Uranus, and Neptune
   b. Uranus, Neptune, and Pluto   **d.** Mercury, Venus, and Saturn

Review (p. 462)
Now that you've finished Section 1, review what you learned by
answering the Review questions in your ScienceLog.

Section 2: Moons (p. 463)

1. What is the difference between a moon and a satellite?

_____

_____

2. Which planets do not have moons?

   a. Mercury and Venus   **c.** Uranus and Neptune
   b. Neptune and Pluto   **d.** Mars and Pluto

Luna: The Moon of Earth (p. 463)

Mark each of the following statements *True* or *False*.

3. _____ The moon's composition is similar to the composition of the Earth's crust.

4. _____ The ages of rocks brought back from the moon during the 1960s and 1970s were measured using radiometric dating techniques.

5. _____ The missions to the moon have not given us any information about other parts of the solar system.

6. Before the Apollo missions, which of the following ideas was NOT used to explain the formation of the moon?
   a. It was captured by the Earth's gravity.
   b. It resulted when the Earth was hit by a planet-sized object.
   c. It spun off from the Earth.
   d. It formed independently from the same materials that make up the Earth.

7. Which of the following is NOT part of the current theory about the formation of the moon?
   a. It resulted when the Earth was hit by another object.
   b. It was formed partially from material from Earth.
   c. Some of the craters on the moon were formed during a cooling period.
   d. It formed independently from materials different from those that make up the Earth.

8. As a result of the changing positions of the moon relative to the Earth and sun, the moon has different appearances, called

   _____ .

9. Take a look at Figure 23. What is the difference between a new moon and a full moon?

   _____

   _____

   _____

10. When the sunlit portion of the moon appears to grow larger, we

    say the moon is _____ .
    (waxing or waning)

11. As the moon changes its position relative to the Earth and sun,

    we always see the same side of the moon. True or False?
    (Circle one.)

_____

Chapter 18, continued

_____

Choose the word in Column B that best matches the definition in Column A, and write the corresponding letter in the space provided.

Column A	Column B
____ **12.** The shadow of the Earth falls on the moon.	**a.** annular eclipse
____ **13.** The shadow of the moon falls on the Earth.	**b.** lunar eclipse
____ **14.** A thin ring is visible around the outer edge of the moon.	**c.** solar eclipse

15. During a total lunar eclipse, the moon often turns deep blue.

   True or False? (Circle one.)

Review (p. 467)

Now that you've finished the first part of Section 2, review what you learned by answering the Review questions in your ScienceLog.

The Moons of Other Planets (p. 468)

16. Phobos and Deimos are moons of Mars. Where do scientists think these moons come from, and why?

_____

_____

_____

_____

_____

17. Ganymede is the largest of Jupiter's 16 moons. It is larger than

   a. Mercury.   **c.** Earth.
   b. Venus.   **d.** Mars.

18. Europa is another of Jupiter's moons. Why do scientists wonder if some form of life might be there?

_____

_____

_____

▲▲ CHAPTER 18
▲▲
▲

Complete the following section after reading pages 469–470. Write the number of moons of each planet in the space provided.

19. _____ Neptune

20. _____ Saturn

21. _____ Uranus

22. _____ Pluto

Choose the moon in Column B that best matches the description in Column A, and write the corresponding letter in the space provided.

Column A	Column B
_____ **23.** revolves backward around its planet	**a.** Titan
_____ **24.** atmosphere is a primordial soup of hydrocarbons and nitrogen	**b.** Charon
_____ **25.** patchwork surface of plains, grooves, and cliffs	**c.** Miranda
_____ **26.** period of revolution equals period of rotation	**d.** Triton

Review (p. 470)

Now that you've finished Section 2, review what you learned by answering the Review questions in your ScienceLog.

Section 3: Small Bodies in the Solar System (p. 471)

1. Name two objects in the solar system besides moons and planets.

_____

_____

Comets (p. 471)

2. What are "dirty snowballs," and why are they given that name?

_____

_____

3. Comets can have more than one tail. True or False? (Circle one.)

4. When comets are at a point in their elliptical orbit closest to the sun, they are at _____, and when they are farthest from the sun, they are at _____.

5. Where do you find orbiting comets? (Circle all that apply.)

    a. the Oort Cloud   **c.** the asteroid belt

    b. the Kuiper Belt   **d.** outside Neptune's orbit

Asteroids (p. 473)

6. Where would you most likely find asteroids?

_____

_____

_____

7. Asteroids have a variety of sizes, shapes, and compositions.

True or False? (Circle one.)

8. Asteroids closest to the sun are

   a. rich in carbon.
   b. stony or metallic in composition.
   c. rich in organic matter.
   d. dark gray on their surfaces.

9. What is the full name of NASA's first mission to study the asteroids?

_____

_____

Meteoroids (p. 474)

10. What is the main difference between a meteoroid and an asteroid?

_____

_____

Mark each of the following statements *True* or *False*.

11. _____ A meteorite is a meteoroid that enters Earth's atmosphere and strikes the ground.

12. _____ A meteor shower occurs when Earth passes through the dusty debris left behind by a comet.

13. _____ All meteorites are made up of the same types of materials.

14. _____ A meteor is a small, rocky body orbiting the sun.

15. Which type of meteorite may contain organic minerals and water?

   a. stony
   b. metallic
   c. stony iron

CHAPTER 18 ▲ ▲ ▲ ▲

Role of Impacts in the Solar System (p. 474)

16. Why does the Earth receive fewer impacts than the moon?

_____

_____

_____

_____

_____

_____

17. Why don't we see many craters on Earth? (Circle all that apply.)

    a. weathering   **c.** few objects hit the Earth
    b. tectonic activity   **d.** no longer happens

18. Many scientists believe that a cosmic impact caused the extinction of the dinosaurs. How often do these kinds of impacts occur?

    a. every few hundred years
    b. every few thousand years
    c. every few million years
    d. every 30 million to 50 million years

Review (p. 475)

Now that you've finished Section 3, review what you learned by answering the Review questions in your ScienceLog.

CHAPTER

19 **DIRECTED READING WORKSHEET**

The Universe Beyond

As you read Chapter 19, which begins on page 482 of your textbook, answer the following questions.

Imagine (p. 482)

1. Robert Williams decided to focus the Hubble Space Telescope on what looked like an empty piece of sky. What did he find?

_____

_____

2. You will learn about _____ and the

_____ they are made of in this chapter.

What Do You Think? (p. 483)

Answer these questions in your ScienceLog now. Then later, you'll have a chance to revise your answers based on what you've learned.

Investigate! (p. 483)

3. Each galaxy contains _____ of stars. (millions or billions)

4. All galaxies are the same shape. True or False? (Circle one.)

Section 1: Stars (p. 484)

5. What is a star?

_____

_____

6. To learn more about stars, astronomers study

_____ .

Color of Stars (p. 484)

7. Place these colors of flame in order from hottest to coolest: red, blue, and yellow.

_____

8. Astronomers think that stars have different

_____ because they are different colors.

CHAPTER 19

Chapter 19, continued

Composition of Stars (p. 484)

9. A prism spreads sunlight into its colors. Similarly, a

_____ spreads starlight into its colors.

Mark each of the following statements *True* or *False*.

10. _____ The part of a star we see gives off a continuous spectrum.

11. _____ Every element has its own set of emission lines.

12. A cool gas _____ the same colors of light

that it would _____ if it were heated.

13. Take a moment to look at the Physical Science Connection in the right column of page 485. How do police use spectrographs to identify cars?

_____

_____

_____

_____

14. A spectrum of starlight taken with a spectrograph is a

_____ spectrum with

_____ lines called an absorption

spectrum.

Classifying Stars (p. 486)

15. Today, stars are classified by temperature. The hottest stars are

    a. yellow.   **c.** red.
    b. orange.   **d.** blue.

16. If a star's spectrum does not contain an absorption line for an element, then the star cannot contain the element.

True or False? (Circle one.)

17. Look at the Life Science Connection in the left column of page 486. Why is it hard to distinguish star colors at night?

_____

_____

_____

Look at the chart of the types of stars on page 487 to answer the following questions.

18. Class F stars are

    a. yellow.   **c.** yellow-white.

    b. blue-white.   **d.** orange.

19. Which element is found in the spectra of the hottest stars?

    a. calcium   **c.** iron

    b. hydrogen   **d.** helium

20. The surface temperature of our sun is

between _____ and

_____ °C.

How Bright Is That Star? (p. 487)

Mark each of the following statements *True* or *False*.

21. _____ The stars we see are different from those seen by ancient astronomers.

22. _____ A first-magnitude star is brighter than a sixth magnitude star.

23. _____ Apparent magnitude is how bright a star looks in the night sky.

24. The sun's absolute magnitude is 4.8, which is not very high. Why is this star the brightest object in the sky?

_____

_____

Distance to the Stars (p. 489)

25. Light travels about _____ km in one light-year.

26. As Earth revolves around the sun, more-distant stars seem to shift position in relation to stars near the Earth. True or False? (Circle one.)

Motions of Stars (p. 489)

27. At different times of the year, the night side of the Earth faces a different part of the universe. True or False? (Circle one.)

28. Some stars seem to _____ and

_____ because of the rotation of the Earth on its axis.

29. Look at Figure 8. If you could travel through time, would the constellations look the same from Earth 200,000 years in the future as they do now? Explain.

_____

_____

_____

Review (p. 490)

Now that you've finished Section 1, review what you learned by answering the Review questions in your ScienceLog.

Section 2: The Life Cycle of Stars (p. 491)

1. When a star dies, either gradually or in a big explosion, much of

its material returns to _____ .

The Diagram That Did It! (p. 491)

2. An H-R diagram shows the relationships of which of the following? (Circle all that apply.)

   a. a star's absolute magnitude
   b. a star's apparent magnitude
   c. a star's surface temperature
   d. a star's velocity

3. An H-R diagram shows how stars can be classified by temperature

and brightness. True or False? (Circle one.)

The H-R Diagram (p. 492)

4. According to the H-R diagram in your textbook, the star Canopus

has a temperature of about _____ and

an absolute magnitude of _____ .

5. Which type of star would you find in the lower right corner of an H-R diagram?

   a. a red, faint star
   b. a blue, faint star
   c. a cool, bright star
   d. a yellow, bright star

6. Main sequence stars move _____ and

to the _____ on an H-R diagram as they

age. (up or down, right or left)

Chapter 19, continued

Choose the type of star in Column B that best fits the description in Column A, and write the appropriate letter in the space provided.

Column A	Column B
_____ **7.** stars with low mass, temperature, and absolute magnitude	**a.** massive blue stars
_____ **8.** small, hot stars that are dimmer than the sun	**b.** white-dwarf stars
_____ **9.** high-temperature stars that are brighter than the sun and that quickly use up their hydrogen	**c.** red giant stars
	d. red-dwarf stars
_____ **10.** cool stars with high absolute magnitudes	**e.** main sequence stars
_____ **11.** stars that are in the band that runs along the middle of the H-R diagram	

When Stars Get Old (p. 494)

12. Which of the following statements is NOT true of supernovas?

   a. The explosion can be brighter than a galaxy.
   b. The explosion occurs at the beginning of a blue star's life.
   c. Silver, gold, and lead can be produced during the explosion.
   d. After the explosion, the supernova may shine for many days.

13. A supernova explosion was observed on Earth in 1987, but the explosion actually took place about 169,000 years ago. How is this possible?

_____

_____

_____

14. Which of the following statements are true of neutron stars? (Circle all that apply.)

   a. They are only about 20 km across.
   b. They are made up of neutrons.
   c. They are not very dense.
   d. They are the leftover material in the center of a supernova.

15. A pulsar is a neutron star that is spinning. True or False? (Circle one.)

16. Which of the following are true of black holes?
(Circle all that apply.)

    a. They are larger than neutron stars.

    b. Light cannot escape the strength of their gravity.

    c. They can gobble up other stars.

    d. Astronomers use X rays to detect their location.

Review (p. 495)

Now that you've finished Section 2, review what you learned by answering the Review questions in your ScienceLog.

Section 3: Galaxies (p. 496)

1. What holds groups of stars, such as galaxies, together?

_____

_____

Types of Galaxies (p. 496)

Each of the following statements is true of a spiral galaxy, an elliptical galaxy, or an irregular galaxy. In the space provided, write *S* for a spiral galaxy, *E* for an elliptical galaxy, and *I* for an irregular galaxy.

2. _____ These galaxies contain only old stars.

3. _____ The Milky Way is probably this type of galaxy.

4. _____ Many of these galaxies may have their gravity distorted by neighboring galaxies.

5. _____ Most galaxies are this type of galaxy.

6. _____ These galaxies are massive blobs of stars.

7. In a spiral galaxy, hot blue stars are located in the

_____ . (spiral arms or nuclear bulge)

Contents of Galaxies (p. 498)

8. Stars are the largest features in galaxies. True or False? (Circle one.)

9. What do nebulas have to do with new stars?

_____

_____

10. Globular clusters usually contain fewer stars than open clusters.

True or False? (Circle one.)

Origin of Galaxies (p. 499)

11. Galaxies may form when clouds of _____

and _____ collapse. One theory states
that if the cloud that forms a galaxy is rotating quickly, a(n)

_____ galaxy is created. A(n)

_____ galaxy forms if the cloud is not
rotating fast.

12. All of the following are true about quasars EXCEPT

   a. quasars are very large.
   b. quasars may be the earliest types of galaxies ever formed.
   c. quasars are very far away.
   d. quasars may have enormous black holes at their center.

Review (p. 499)

Now that you've finished Section 3, review what you learned by
answering the Review questions in your ScienceLog.

Section 4: Formation of the Universe (p. 500)

1. The study of the _____ and

_____ of the universe is cosmology.

The Big Bang Theory (p. 500)

2. According to the big bang theory, the universe began 10 billion to

15 billion years ago with a huge explosion. True or False?
(Circle one.)

3. If there is enough matter in the universe, it may stop expanding

outward and start collapsing. True or False? (Circle one.)

4. Which of the following is evidence that supports the big bang
theory?

   a. stars getting older and becoming dark
   b. background radiation coming from all directions in space
   c. excess amounts of mass in the universe
   d. a large vacuum in space

Universal Expansion (p. 501)

Mark each of the following statements *True* or *False*.

5. _____ Scientists have shown that the Milky Way is near the
center of the universe.

6. _____ Almost all of the galaxies in the universe are moving
away from all of the other galaxies in the universe.

CHAPTER 19

7. To discover the age of the universe, what do scientists have to know about galaxies?

_____

_____

_____

8. Another way to measure the age of the universe would be to measure the age of the oldest stars in the universe. What has been the problem with this method so far?

   a. Scientists have not found any stars that are very old.
   b. None of the very old stars are close enough to Earth to be accurately measured.
   c. The measurements show that some stars are older than the universe.
   d. The sun is the only star in our galaxy to measure.

Structure of the Universe (p. 503)

Arrange the following structures in order from smallest to largest, with 1 being the smallest, by writing the appropriate number in the space provided.

9. _____ galaxies

10. _____ planets

11. _____ superclusters

12. _____ star clusters

13. _____ galaxy groups

14. _____ planetary systems

15. _____ galaxy clusters

Review (p. 503)

Now that you've finished Section 4, review what you learned by answering the Review questions in your ScienceLog.